Milton and Newton

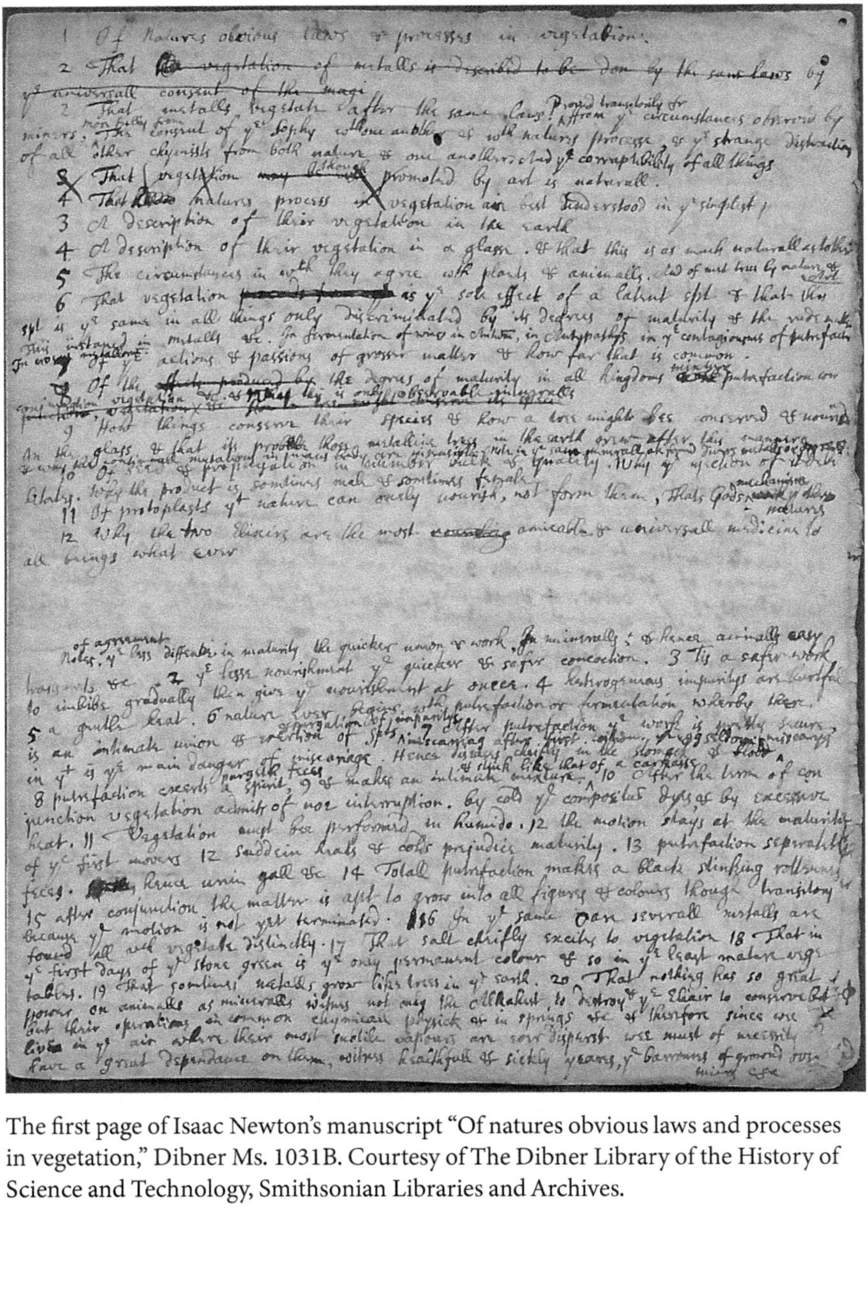

The first page of Isaac Newton's manuscript "Of natures obvious laws and processes in vegetation," Dibner Ms. 1031B. Courtesy of The Dibner Library of the History of Science and Technology, Smithsonian Libraries and Archives.

Milton and Newton

Matter, God, and Modernity

STEPHEN M. FALLON

OXFORD
UNIVERSITY PRESS

OXFORD
UNIVERSITY PRESS

Great Clarendon Street, Oxford, OX2 6DP,
United Kingdom

Oxford University Press is a department of the University of Oxford.
It furthers the University's objective of excellence in research, scholarship,
and education by publishing worldwide. Oxford is a registered trade mark of
Oxford University Press in the UK and in certain other countries.

Published in the United States of America by Oxford University Press
546 Fifth Avenue, New York, NY 10036, United States of America

British Library Cataloguing in Publication Data
Data available

Library of Congress Control Number: 2026945972

ISBN 9780198988908

DOI: 10.1093/9780198988939.001.0001

Printed and bound by
CPI Group (UK) Ltd., Croydon, CR0 4YY

The manufacturer's authorized representative in the EU for product safety is
Oxford University Press España S.A. of Parque Empresarial San Fernando de Henares,
Avenida de Castilla, 2 – 28830 Madrid (www.oup.es/en or product.safety@oup.com).
OUP España S.A. also acts as importer into Spain of products made by the manufacturer.

MIX
Paper | Supporting
responsible forestry
FSC™ C013604
FSC
www.fsc.org

This Book is Dedicated
to
Sanford Budick and James Nohrnberg
and to
My Teachers, Colleagues, and Students,
in All of Whom I Have Been Fortunate

A short time before his death [Newton] uttered this memorable sentiment: "I do not know what I may appear to the world, but to myself I seem to have been only like a boy playing on the sea-shore, and diverting myself in now and then finding a smoother pebble or a prettier shell than ordinary, whilst the great ocean of truth lay all undiscovered before me."[1]

collecting toys,
And trifles for choice matters, worth a sponge;
As children gathering pebbles on the shore. (*Paradise Regained* 4.328–30)

A friend once said to him, "Sir Isaac, what is your opinion of poetry?"—His answer was; "I'll tell you that of Barrow;—he said, that poetry was a kind of ingenious nonsense."[2]

[1] As recorded by David Brewster, *Memoirs of the Life, Writings, and Discoveries of Sir Isaac Newton*, vol. 2 (Edinburgh: T. Constable and Co., 1855), 407. http://www.newtonproject.ox.ac.uk/view/texts/normalized/OTHE00088, accessed July 8, 2020.

[2] Joseph Spence, *Anecdotes, Observations, and Characters, of Books and Men: Collected from the Conversation of Mr. Pope, and Other Eminent Persons of His Time*, ed. Samuel Singer (London: W. H. Carpenter, 1820), 368.

Contents

Preface

When John Milton, England's greatest narrative poet, died on November 8, 1674, Isaac Newton, England's greatest natural philosopher, was two months shy of his thirty-second birthday, and their friend and intellectual networks overlapped. In this book, I argue that attention to the surprising and arresting parallels between their speculations on the life of matter, their antitrinitarianism, and their self-understanding as prophets can provide a glimpse into the ferment of late seventeenth-century thought and the expansion of what was newly possible to think. Exploration of the parallels will also lead to a fuller understanding of Milton's place in intellectual history and the significance of his thought, as well as the motivations and implications of the rich, if less well-known, body of Newton's unpublished natural philosophical and theological manuscripts.

In this study I return to the interests of my two previous books on Milton. In *Milton among the Philosophers*, I placed Milton's distinctive animist materialism in the context of seventeenth-century debates on the relation of matter and spirit. In *Milton's Peculiar Grace*, I explored his paradoxical relation to contemporary Christian theological understandings and practice around grace, sinfulness, and election. In this book, I will suggest that in Milton and Newton similar natural philosophical views are connected to their unusual and, from orthodox perspectives, heretical Christian positions. I will revise my earlier argument for the idiosyncrasy of Milton's understanding of the relation of spirit and body, as I place it now in the context of seventeenth-century vitalist natural philosophy in general and of Newton's speculation on active principles in matter in particular. Throughout I will illustrate that they echo each other in sometimes uncanny ways, and that they diverge in ways dictated by their deepest commitments, Milton's to freedom of the will and Newton's to divine omnipotence.

In the Introduction I provide a brief overview of the convergences in the philosophical and religious thought of Milton and Newton that I will develop in the rest of the book, before taking up—through an outline of large and significant developments in the philosophical, natural philosophical, and theological discourses to which they were compelled to respond—the question of why they might have arrived at similar positions. I then address competing guiding ideas—Milton's to free will and Newton's to divine omnipotence—that account for the key divergences between otherwise similar positions. Finally, I provide an overview of the theological and alchemical thinking of Newton in his voluminous manuscripts, and describe his commitment to recovering the *prisca scientia* and *prisca theologia*, which both drives his obsessive and repetitive work, particularly in the religious manuscripts, and offers a point of difference from Milton.

Part 1 of the book focuses on Milton's and Newton's views on matter. Chapter 1 will examine Milton's and Newton's shared vitalism. For Milton, and for Newton at times, matter is not essentially dead and inert, requiring the addition of incorporeal substance to account for life and motion; instead, matter is itself instinct with life; for both matter is either living or, in Newton's case, ready to be kindled into life without the superaddition of incorporeal substance. Milton's vitalism, or animist materialism, is familiar to Milton scholars, but they may be surprised to find that Newton, demonized by William Blake as a proponent of mechanism, speculated often about the vitality of matter. Even more surprising is Newton's articulation of a cosmic vitalism expressed in language recalling Milton's in *Paradise Lost*. This extraordinary convergence has implications for how we should understand what might otherwise seem Milton's retrograde, fanciful, and merely poetic picture of a living universe.

Chapter 2 turns to alchemy, as providing the language for Milton's expressions of material life, and as the natural philosophical study where Newton focused his attempts to uncover and understand active and vital material agents. Newton, as historians of science have learned as his voluminous manuscript writings have become in recent decades more easily available, aspired to be an alchemical adept. Milton's own vitalist thinking is more explicable the more one understands Early Modern alchemy as a branch of natural philosophy that gave rise to modern chemistry rather than as an occult and magical art. If Milton in his great epic seems dismissive of alchemy, the example of Newton will suggest that skepticism about attempts to fabricate the philosophers' stone does not indicate dismissal of the science of alchemy. Newton's understanding of the cosmos and creation, like Milton's, is alchemical, and medical alchemy will provide a context in which to understand Milton's extraordinary picture of the transformation of material food to material soul and reason in Book 5 of *Paradise Lost*.

In Part 2 I turn to Milton's and Newton's theological and religious thought. Chapter 3 examines Milton's and Newton's parallel self-understandings as prophets in a post-prophetic age. Prophetic election for each is something earned through a pure and chaste life and exercised not passively but through diligent labor in the Word. The prophetic tasks differ in characteristic ways, with Milton producing inspired narratives of unrecorded providential history and Newton scouring recorded history to demonstrate that an omnipotent God has predicted in Scripture, and by extension directed, the unfolding of that history.

In Chapter 4 I compare Milton's and Newton's antitrinitarian, Arian theologies, as well as their parallel attraction to Socinian positions. I will point to their similar methods of biblical interpretation, which led them to their arguments that the Son of God, while begotten before creation, is neither co-essential nor co-eternal with the Father, who is the one God. Newton developed his Arian understanding of the Son, as we shall see, after the publication of *Paradise Lost* and in the last years of Milton's life. Newton scholars, like Milton scholars, have identified in their subject a hybrid of Arian theological views and Socinian religious views. I will look at the incomplete and implicit gestures that each makes in the direction of Socinianism, including the belief that the Son of God was a human being chosen for his merit for promotion to the godhead. The tendency toward Socinianism is connected to a shared aversion to mystery and a shared belief that the central and saving doctrines of Christianity are consistent with reason.

In the Conclusion, I will compare Milton as the poetic visionary and Newton as the natural philosophical visionary. They faced different challenges, dictated by the different discourses of poetry and natural philosophy. Those facing Newton are the more obvious: he had to develop mathematical theories to account for meticulously observed phenomena and, in turn, demonstrate the predictive reliability of those theories. Milton's task seems at first glance significantly easier. As a poet apparently licensed, as Philip Sidney suggested, to "freely range within the zodiac of his own wit," Milton nevertheless committed himself to writing what was to be in his intention a true epic tracing the arc of sacred history against the backdrop of the cosmos as understood in his time. As a result, he took on two intractable and perhaps impossible tasks: (1) placing an immutable God in motion and in conversation with a divine but not co-equal Son, and (2) creating a stage that followed the laws of his own vitalist and materialist understanding of substance and putting into play on that stage the spectrum of beings from God and angels to objects conventionally assumed to be inanimate.

The fact that I come to the subject of this book as a Milton scholar will be apparent to readers. As I have studied Newton's works and consulted the work of Newton scholars, I have focused on those elements of Newton's writings, and particularly of his unpublished manuscripts, that reveal affinities between these important figures. Part of my aim in writing this book is to show that Milton's picture of the cosmos and his approach to the life of matter in *Paradise Lost* should be regarded as more than mere poetic flights of fancy and that, moreover, they are interconnected with his theological speculations in a way comparable to Newton's attempt to assimilate his natural philosophical, cosmological, and theological speculations into a unified body of thought.

Acknowledgments

My exploration of the relation of John Milton and Isaac Newton was prompted by a 2009 conference at the University of Sussex, "Milton/Newton: Two Cultures," which brought together literature scholars and historians of science. At the time, I did not anticipate that the relation between the poet's and natural philosopher's thought would become the focus of years of reading culminating in this book. Thus, I owe thanks to the conference organizers who invited me: Brian Cummings and Rob Iliffe.

I have accrued many debts over my years working on this project. I owe the greatest thanks to Timothy Harrison, who has read more than one version of the manuscript and whose suggestions and challenges have made the book better. Sanford Budick, Henry Weinfield, and John Sitter have read versions of the whole, and each has contributed to whatever is valuable in it. My old friend John Rumrich, who was the first to point Milton scholars to Newton, has been an invaluable interlocutor. For advice and responses, I am indebted to John Rogers and Rachel Trubowitz, both of whom have advanced our understanding of the relation of Milton and Newton. For conversations about Newton's vitalism, I owe thanks to Lawrence Principe and William Newman. Colleagues at the Northeast Milton Seminar, the Canada Milton Seminar, and the Newberry Milton Seminar led me to sharpen my argument and avoid missteps. A long but still partial list of scholars from whose writing and advice I have benefited includes Sharon Achinstein, David Aers, Sarah Beckwith, Gregory Chaplin, Alison Chapman, Ann Coiro, Dennis Danielson, Amrita Dhar, Stephen Dobranski, Karen Edwards, Philippa Earle, Noam Flinker, Jeffrey Gore, Achsah Guibbory, Dayton Haskin, Stephen Hequembourg, Edward Jones, Christopher Kendrick, Maggie Kilgour, William Kolbrener, Ayelet Langer, Russ Leo, John Leonard, David Loewenstein, Catherine Gimelli Martin, Jeffrey Miller, William Miller, Ryan Netzley, Charlotte Nicholls, Mary Nyquist, Kristen Poole, Jason Rosenblatt, Regina Schwartz, Joshua Scodel, Daniel Shore, Lauren Shohet, William Shullenberger, Nigel Smith, Paul Stevens, Richard Strier, Noel Sugimura, Pasquale Toscano, and Tzachi Zamir. Two dear friends sadly no longer with us have left their marks on this book: Emily Budick and Michael Lieb. Notre Dame colleagues whom I want to thank for their support and their responses to the work in progress include Laura Knoppers, Clark Power, Matthew Rickard, and Elliot Visconsi. My graduate students have been a worthy first audience for much of what appears here, and I want to record my gratitude to Sarah Baber, Dong Hwan Chun, Ala Fink, Ethan Guagliardo, Jay David Miller, and Arnaud Zimmern. I am indebted to the anonymous readers for Oxford University Press; their advice led to a sharper framing of the study.

I am grateful to lecture audiences at Princeton University, the University of Toronto, Duke University, The Hebrew University of Jerusalem, the University of Rochester, the Twelfth International Milton Symposium in Strasbourg, the Universidade Federal de Minas Gerais (Brazil), and Zhejiang University (China).

Two chapters draw from previously published essays, Chapter 1 from "John Milton, Isaac Newton, and the Life of Matter," which appeared in *Milton and the New Scientific Age* (Routledge, 2019), and Chapter 4 from "Milton, Newton, and the Implications of Arianism," which appeared in *Milton in the Long Restoration* (Oxford University Press, 2016). I am grateful to the publishers for permission to use this material. The frontispiece image appears courtesy of The Dibner Library of the History of Science and Technology, Smithsonian Libraries and Archives.

Research for this book was supported by a 2016 Guggenheim Fellowship. The Franco Institute at the University of Notre Dame supported the work of the indexer.

I have been fortunate in my editors at Oxford University Press, Eleanor Collins and Jacqueline Norton, and in the production team, led by Jo Spillane and Srividya Raamadhurai.

Here, as I turn from professional to personal debts, I want to thank my son Sam, to whom I am indebted in both registers. Special thanks to Joan Wulff Fallon, with whom I share "Collateral love, and dearest amity." I want to thank my children and their partners for listening patiently as I shared more than they needed to know about what is in this book: Sam and Anne, Claire and Greg, Andy and Cam, Dan and Megan, and Martin. The past seven years I have been buoyed by the arrival of, and the joy of time with, our grandchildren, Maggie, Max, Gabe, Noah, and Arjun.

Abbreviations

CPW *The Complete Prose Works of John Milton*, edited by Don M. Wolfe et al., 8 vols. (New Haven, 1953–82).

DDC Milton, *De Doctrina Christiana*

MLM Modern Library Milton, *The Complete Poetry and Essential Prose of John Milton*, edited by William Kerrigan, John Rumrich, and Stephen M. Fallon (New York: Modern Library, 2007).

OCW *The Complete Works of John Milton,* Vol. 8, Parts 1 and 2 (*De Doctrina Christiana*), ed. And trans. John K. Hale and J. Donald Cullington (Oxford: Oxford University Press, 2012).

My citations of the Bible, as opposed to those in quotations, are taken from the King James Version.

Introduction

Studies of literature and the history of science often take one of two paths, either applying the tools of literary analysis to natural philosophers' and scientists' texts or reading literary texts in the light of contemporary natural philosophical and scientific developments. My approach will be different, as I will look at a leading poet and natural philosopher side by side and ask why they responded similarly to newly urgent questions: with the decline of the Aristotelian hylomorphic understanding of substance, and in a context of contending dualist and mechanist materialist philosophies, how should one think of body, mind, and soul, and what is the definition and source of life? With the advent of the Higher Criticism of the Bible and the emphasis on a return to primitive Christianity, how should one think about the consensus on the doctrine of the Trinity promulgated in the fourth century's Nicene Creed? Milton and Newton lived in a world in which the understanding of the physical universe was rapidly and fundamentally changing, and the changes can help us to understand how and why Milton and Newton might have arrived at similar vitalisms, similar speculations on the intrinsic vital power of the body, and similar skepticism on the immortality of the soul. Milton and Newton, two formidably learned and curious thinkers, responded to seventeenth-century intellectual revolutions that rivaled in intensity and significance their country's political revolutions, and their thought contained seeds of developments in the centuries to follow.

At first glance, the things that separate John Milton and Isaac Newton seem to overwhelm the things that connect them. A poet and bold re-imaginer of biblical myth and history stands on one side, and on the other a mathematician and experimentalist natural philosopher reportedly dismissive of poetry. One is credited with summing up and crowning the knowledge and literary forms of preceding centuries and millennia, and the other with revolutionizing natural philosophy and ushering in a new way of thinking about the cosmos and the relation of physical bodies that would dominate centuries to come. Milton, who never held public office, offered running commentaries on the great public questions of his day; though twice Member for Cambridge, Newton entered political controversy only cautiously (according to a telling anecdote, he spoke only once in Parliament, to ask an usher to close a window).[1] Newton published reluctantly,

[1] Richard Westfall, *Never at Rest: A Biography of Isaac Newton* (Cambridge: Cambridge University Press, 1980), 288. Newton, however, alarmed at what he saw a move to undermine the Protestant nature of Cambridge University, did take an active role in resisting James II's attempt to mandate that

fearing that his time would be wasted and equanimity disrupted by the correspondence and controversy that publication would invite. Milton published regularly and eagerly, never hesitating to court controversy, and at the end of his life scouring his papers for works to be sent to the printers. First glances, however, can be deceptive. Milton had a strong amateur's interest in mathematics, and the mathematically informed works of Rachel Trubowitz and Erin Webster have connected the imagery of *Paradise Lost* with the infinitesimals of Newton's calculus.[2] In this book I trace arresting and significant convergences in the integrated philosophical and theological views of these two intellectual giants, and I will argue that their otherwise parallel constellations of thought are colored by differing central commitments, Milton's to freedom of the will and Newton's to divine omnipotence.

Some perspectives shared by Milton and Newton are not surprising for English Protestants in the mid- and late seventeenth century. From comfortable middle-class backgrounds, both were raised in the Anglican church and took their degrees at Cambridge.[3] Both were conventionally fervent anti-Catholics, an animus that may in Milton's case have been magnified by his Catholic grandfather's disowning his Protestant father. They shared the Protestant emphasis on *sola scriptura*, working out one's faith for oneself by reading and interpreting the Bible, though, as we shall see, they followed this imperative into doctrinal unorthodoxies not typical of nearly all of their contemporaries.

Both contrasted the intellectual freedom possible in Protestant England with the restrictions endured by intellectuals in Catholic Italy. Milton recorded having heard from "lerned men" during his fifteen-month Italian tour (Spring 1638 to Summer 1639) that he was "happy to be born in such a place of *Philosophic* freedom, as they suppos'd English was, while themselves did nothing but bemoan the servil condition into which lerning amongst them was brought.... There it was that I... visited the famous *Galileo* grown old, a prisner to the Inquisition, for thinking in Astronomy otherwise then the Franciscan and Dominican licensers

Alban Francis, a Benedictine monk, be admitted as Master of Arts without taking the Oath of Supremacy (Westfall, *Never at Rest*, 474–79; Rob Iliffe, *Priest of Nature: The Religious Worlds of Isaac Newton* [Oxford: Oxford University Press, 2017], 295–98).

[2] Rachel Trubowitz, "Reading Milton and Newton in the Radical Reformation: Poetry, Mathematics, and Religion," *ELH* 84 (2017): 33–62; Erin Webster, "Milton's Pandæmonium and the Infinitesimal Calculus," *English Literary Renaissance* 45 (2015): 425–58, a revised version of which appears as Chapter 5, "Perspective as a Perceptual Tool in Milton and Newton," in Erin Webster's *The Curious Eye: Optics and Imaginative Literature in Seventeenth-Century England* (Oxford: Oxford University Press, 2020).

[3] The standard source for Newton's life, upon which I draw throughout, is Westfall's *Never at Rest*. For Milton's early life, see especially Barbara Lewalski, *The Life of John Milton: A Critical Biography* (Oxford: Blackwell, 2000); Gordon Campbell and Thomas N. Corns, *John Milton: Life, Work, and Thought* (Oxford: Oxford University Press, 2010); and Nicholas McDowell, *Poet of Revolution: The Making of John Milton* (Princeton: Princeton University Press, 2020). Campbell and Corns differ from Lewalski and from McDowell in placing Milton's radicalism later in his life.

thought."[4] John Conduitt, Newton's confidante and the husband of Newton's niece, recalls Newton speaking of his gratitude for "being born in a land of liberty & in an age where he {might} speak his mind—not afraid of {the} Inquisition as Galileo was for {saying} the sun stood still & the earth {moved}."[5]

If Newton like Milton prized the intellectual freedom honored, at least in theory, in Protestant England, he also shared Milton's sense of the responsibility of individuals and peoples to earn and deserve their liberty. As Milton's Raphael in *Paradise Lost* explains to a despondent Adam outraged at the vision of Nimrod's tyranny, when man "permits / Within himself unworthy powers to reign / Over free reason, God in judgment just / Subjects him from without to violent lords" (*Paradise Lost* 12.90–93).[6] In *Animadversions*, Milton offers a succinct description of his political ideal when he calls the Parliamentary side in the impending Civil War "a Kingdome of free spirits" (*CPW* 1:669). Virtuous individuals, by governing themselves within, are worthy of governing themselves politically. Being placed under arbitrary rule is the consequence of and fitting punishment for failure to rule oneself. Newton tells a similar story in his massive manuscript "Of the Church." The early Christians, Newton writes, followed the two great commandments, to love God and to love their neighbors, "while they continued in affliction under the Roman heathen Emperors," because the suffering winnowed from their ranks "hypocrites and vitious men." Trouble arose along with ease, when the Emperor Constantine lifted the persecution and the Christian community became not a target but a client of the state: "the Christians by the prosperity of their affairs...& by the crowding in of flatterers & ambitious men into Ecclesiastical dignities,....they were grown more vicious in all respects then the northern heathens who invaded them & therefore God was just in making the heathens lords over them."[7] In Milton as in Newton, subjugation to rulers is a just divine punishment for vice and failure in self-government, and in both again this does not excuse the tyrant.[8]

[4] *Areopagitica*, in the *Complete Prose Works of John Milton*, ed. Don M. Wolfe et al., 8 vols (New Haven: Yale Univer, 1953–82), 2: 537–38. Further references to Milton's prose, with the exception of *De Doctrina Christiana*, will be taken from this edition, and cited as *CPW*.

[5] John Conduitt, "Notes on Newton's Character," Keynes Ms. 130.07, 5^r; http://www.newtonproject.ox.ac.uk/view/texts/normalized/THEM00169, accessed August 6, 2025.

[6] See similar passages in *Samson Agonistes* 268–71, "But what more oft in nations grown corrupt, / And by their vices brought to servitude, / Than to love bondage more than liberty, / Bondage with ease than strenuous liberty," and *Paradise Regained* 2.466–72. Here and throughout, I quote Milton's poetry from *The Complete Poetry and Essential Prose of John Milton*, ed. William Kerrigan, John Rumrich, and Stephen M. Fallon (New York: Modern Library, 2007), cited hereafter as *MLM*.

[7] Newton, "Of the Church," Chapter 7, (fols. 3–4). This large manuscript—paginated separately by chapter, not reproduced in The Newton Project, and yet to be transcribed—is housed at the Fondation Martin Bodmer, Geneva, Switzerland. I am grateful to the Bodmer Foundation for supplying a digital copy of the manuscript.

[8] Milton shares Newton's harshly critical sense of the corrupting effect of the Donation of Constantine; see, e.g., *Of Reformation* (*CPW* 1:558–60).

Milton, as is well-known, was prominent in the opposition to what he viewed as the arbitrary rule and tyranny of Charles I, laying out the case for the deposing of tyrants in *The Tenure of Kings and Magistrates* and defending the 1649 execution of the king to the English in *Eikonoklastes* (1649) and to Europe in the Latin *Defences of the English People* (1651 and 1654). Three decades later, Newton opposed what he saw as the arbitrary rule and tyranny of James II, denied the divine right of kings, and was closely associated with radical Whigs such as John Hampden, who participated in the Rye House Plot to assassinate James II's father and Charles I's son, Charles II.[9]

While their views on the Roman church's constraints on intellectual freedom and on the political subjugation of those who have enthralled themselves internally are interesting in themselves, I focus in this book on positions that set them apart from the body of their contemporaries and compatriots, as I take up John Rumrich's prescient observation of three decades ago, that although "Milton scholars tend to ignore him, [Newton's] history brings out salient aspects of Milton's life and art better than many of the Protestant divines that we regularly adduce to set a context."[10] I will explore idiosyncratic positions that they share: speculations on the life of matter, claiming the role of inspired prophets, and vigorous denial of the co-equal and co-essential divinity of the Son of God.

The similarities between their bodies of thought have not come into focus before now for four reasons. First, owing to the siloing of academic disciplines scholars interested in one have until recent decades not been concerned with the other. Second, the areas in which Milton's and Newton's thinking overlaps have been seen as at best incidental to and at worst embarrassing for their reputations: their theological unorthodoxy and their alchemically-based speculations on the life, actual or potential, of all matter. Third, despite their three decades as contemporaries, Milton has often been categorized as the last great figure of the English Renaissance and Newton as the first great figure of the English Enlightenment.[11] (It is the case that Newton's 1687 *Principia Mathematica*, by the author's design obscure to all but the most mathematically adept, did not have significant influence until it was popularized at the very end of the seventeenth and the beginning of the eighteenth centuries.[12]) And fourth, the Newton whose natural philosophy and theology provide clues to the coherence and contemporary relevance of Milton's is to be found in a voluminous body of unpublished alchemical and religious manuscripts, described later in this introduction, which have only relatively recently been widely available to scholars.

[9] Iliffe, *Priest of Nature*, 355–62.

[10] John Rumrich, *Milton Unbound: Controversy and Interpretation* (Cambridge: Cambridge University Press, 1996), 37.

[11] Increasingly, scholars are focusing on Milton's relation to post-Restoration and early eighteenth-century contexts; a salient example is the collection *Milton in the Long Restoration*, ed. Blair Hoxby and Ann Baynes Coiro (Oxford: Oxford University Press, 2016).

[12] Stephen D. Snobelen, "On Reading Isaac Newton's *Principia* in the Eighteenth Century," *Endeavour* 22, no. 4 (1998): 159–63.

The unpublished Newton manuscripts clearly reveal intertwined speculations on the nature of matter and on the nature of the Christian godhead that belong to a unified program of thought and that parallel a similar unified program of thought in Milton. In both cases, as I will argue in this book, these programs of thought respond to fundamental intellectual revolutions in the mid-seventeenth century. The unity of Newton's thought in particular becomes evident when "historians," as Stephen Snobelen writes, "[i]nstead of imposing foreign, modern sensibilities of science on early modern natural philosophy,... reconstruct the thought-world and motivations of early modern natural philosophers, who inhabited a world markedly different from our own."[13] If we think of science as predominantly secular, then we might struggle to understand why the founder of modern physics spent so much time on biblical prophecy and the history of Christianity, and we might be inclined to consider Newton's religious thought as a sideline with little or no connection to his thinking about matter and its properties. But it is anachronistic to think of Newton as a scientist; he lived and worked before our conception of science obtained. Newton was a natural philosopher, and, for him as for many of his contemporaries, God's role in the ordering of nature and ordaining final causes were important parts of natural philosophy.[14] In his recent monumental study of a sea change in seventeenth-century philosophy and theology, Dmitri Levitin writes, "If we are to understand the intellectual transformation of early modernity,... we must try to capture as fully as possible what mattered to our actors themselves."[15]

What mattered to Milton and to Newton? As a young man Milton shared his desire to one day write a poem "doctrinal and exemplary to a nation."[16] *Paradise Lost* is doctrinal in two senses, theological and natural philosophical. In *Paradise Lost*, he attempts to "justify the ways of God to men" by giving narrative form to the argument he makes in his Latin *De Doctrina Christiana* (*Christian Doctrine*), that the Fall in Eden was freely willed rather than predestined or ordained by God; if Adam and Eve were not free, then one must conclude that God willed their sin. And *Paradise Lost* is a Lucretian epic as well as a biblical epic; it is an epic of the "way things are." In his effort to write an epic both encyclopedic and true, Milton goes out of his way to construct a vision of a universe animate and

[13] Stephen D. Snobelen, "To Discourse of God: Isaac Newton's Heterodox Theology and His Natural Philosophy," in *Science and Dissent in England, 1688–1945*, ed. Paul Wood (London: Routledge, 2004), 42.

[14] Snobelen, "To Discourse of God," 39–46. Snobelen quotes (43) the historian of science Richard Popkin's only half-joking comment that we should not ask, "'why one of the world's greatest scientists should have spent so much time thinking and writing about religious matters', but 'why did one of the greatest anti-Trinitarian theologians of the 17th century take time off to write works on natural science, like the Principia Mathematica?'" ("Newton's Biblical Theology and His Theological Physics," in *Newton's Scientific and Philosophical Legacy*, ed. P. B. Scheuer and G. Debrock [Dordrecht: Kluwer, 1988], 81).

[15] Dmitri Levitin, *The Kingdom of Darkness: Bayle, Newton, and the Emancipation of the European Mind from Philosophy* (Cambridge: Cambridge University Press, 2022), 4.

[16] Milton, *Reason of Church-Government*, *CPW* 1:815.

free from top to bottom and to deny an ontological division of body and spirit. For Milton as for Newton, natural philosophy could not be divorced from theology, as both addressed the relation of God to the world God created.

Newton directly addressed more than once what mattered to him in his intellectual life, most famously in the General Scholium added to the Second Edition of the *Principia Mathematica* (1713), where he devotes several pages to God's existence and to God's presence in, and dominion over, the universe. Newton argues that "This most elegant system of the sun, planets, and comets could not have arisen without the design and dominion of an intelligent and powerful being," and insists at the end of his discussion, "to treat of God from phenomena is certainly a part of 'natural' philosophy."[17] The recent availability to scholars of Newton's theological writings belies a view, which took hold in the eighteenth century, that an aging, diminished, and even deranged Newton turned to theology. The manuscript record reveals that his interest in theology, like his interest in alchemy, dates from well before the composition of the *Principia*. Newton in 1679 wrote to Robert Hooke suggesting his primary interest may have been something other than mathematical physics:

> I ha[ve] for some years past been endeavouring to bend my self from Philosophy to other studies in so much yt I have long grutched the time spent in yt study unless at idle hours sometimes for a diversion.... [M]y affection to Philosophy being worn out,... I am almost as little concerned about it as one tradesman uses to be about another man's trade or a country man about learning. I must acknowledge my self avers from spending that time in writing about it wch I think I can spend otherwise more to my own content & ye good of others.[18]

How was Newton spending his time otherwise? In the 1670s and 1680s and beyond Newton was busy with theological and alchemical study and manuscripts, including works that will feature largely in this book, such as the "Untitled Treatise on Revelation" of "Of natures obvious laws and processes in vegetation."

The Early Modern Revolt against Metaphysics

That two very different thinkers pursued similar minority speculations on fundamental matters tells us something about what questions might be asked and what might be thought as a result of a fundamental and seismic shift in the early

[17] Newton, *The Principia: Mathematical Principles of Natural Philosophy*, a new translation by I. Bernard Cohen and Anne Whitman, assisted by Julia Budenz, preceded by I. Bernard Cohen, "A Guide to Newton's *Principia*" (Berkeley: University of California Press, 2016), 940, 943.

[18] Newton, Letter to Hooke, November 28, 1679, in *The Correspondence of Isaac Newton, Volume 2: 1676–1687*, ed. H. W. Turnbull (Cambridge: Cambridge University Press, 1959), 300, 302.

modern intellectual world, the reaction against metaphysics. Their sympathy for the reaction against metaphysics is evident, for example, in their shared hostility to the importation into theology, and especially into Christology, of Greek philosophy and Aristotelian metaphysical terms. And in Newton's case it is apparent in the resolution to exclude from the *Principia*'s mathematical model of forces any speculation on the *causes*, whether material or immaterial, of observed physical phenomena. Such speculations, however, are frequent in his unpublished writings, including letters, canceled Prefaces to the *Principia*, and alchemical manuscripts. In the work on which his reputation mainly rests, he avoids being drawn into controversies about the fundamental nature of material things and whether there can be incorporeal causes for physical phenomena. Instead Newton proposes to offer a mathematical model of unspecified forces that account for observable phenomena. In his famous declaration in the General Scholium that concludes the second and third editions of the *Principia*, "*hypotheses non fingo* (I do not feign hypotheses"), Newton insists that he does not claim to identify the cause of gravity:

> I have not as yet been able to deduce from phenomena the reason for these properties of gravity, and I do not feign hypotheses. For whatever is not deduced from phenomena must be called a hypothesis; and hypotheses, whether metaphysical or physical, or based on occult qualities, or mechanical, have no place in experimental philosophy.... [I]t is enough that gravity really exists and acts according to the laws that we have set forth and is sufficient to explain all the motions of the heavenly bodies and of our sea.[19]

This is an intellectually ascetic Newton, hesitant to pronounce or even speculate on the material or immaterial causes of phenomena that he charts mathematically. He adheres to this determination in the *Principia*, with the significant exception, to which I will return later, of the final brief paragraph of the General Scholium and thus of the work as a whole, a paragraph in which he does speculate, however briefly, on those causes.[20] In his manuscripts he strides through the door he leaves slightly ajar here.

What can we make of an apparent tension between Newton's principled agnosticism on the ontological status of gravity and other active principles in the *Principia* and his speculation on that status elsewhere, most notably, as we shall see in Chapter 2, in his alchemical thinking? According to I. Bernard Cohen, "There is no way of telling whether the Newtonian style [in the *Principia*] of initially dealing with the subject on a mathematical rather than a physical plane

[19] Newton, *Principia*, 943.

[20] Here Newton refers to "a certain very subtle spirit" and "electrical bodies" that may account for the attractive and repulsive forces that he describes in the *Principia* (943–44).

was only a subterfuge to avoid criticism or a sincere expression of methodological principle."[21] The answer may be that Newton was doing both, characteristically avoiding controversy that might arise from airing his speculation on the makeup of active principles explored in unpublished manuscripts and at the same time expressing a methodological principle foundational for his mathematical physics. Newton defines rational mechanics in the Preface to the *Principia* as "the science, expressed in exact propositions and demonstrations, of the motions that result *from any forces whatever* and of the forces that are required for any motions whatever," adding that the "causes" of the forces "are not yet known."[22] Newton here leaves open the question of the causes of the forces that he has demonstrated, assigning it to investigation by other sciences, which, once again, he pursued elsewhere.

As Levitin has recently demonstrated, Newton's rational mechanics is best understood in the context of the eclipse of classical Greek metaphysics, a process largely complete by the middle of the seventeenth century.[23] That process ended a long history of metaphysical engagement in Christian apologetics, a discourse devoted both to demonstrating the compatibility of reason and Christian belief and to reaching out to potential converts among the pagans.[24] Patristic authors sought a valid kernel of pagan theology and philosophy, accepted by elites and purportedly derived from contact with ancient Jewish sources (Moses figuring prominently). They suggested that this true pagan wisdom embraced a monotheism lying beneath a polytheistic veneer invented for the vulgar. Church Fathers proposed that unaided human reason gave pagans access to those truths of Christianity not requiring divine revelation, for example, that the soul might somehow "participate in the divine, or... communicate with the eternal logos."[25] Early modern historical and philological research upended the chronology of patristic apologists, as it became clear that any transmission of philosophical ideas, including natural philosophical ideas, went from Egypt to Israel rather than from Israel to Egypt.[26] A new focus on the monist and materialist thought of pre-Socratic philosophers such as Democritus, Leucippus, and Epicurus undermined understandings of ancient philosophy and theology built upon Aristotle's form-matter metaphysics, the Platonic model of the world of forms shadowed by the lower, phenomenal world, and the emanationist Neoplatonic Trinity of One, Mind, and Soul, all three of which were susceptible to reconciliation with Christianity. Now it appeared, as Levitin argues, "the earliest Greek

[21] Cohen, "Guide to Newton's *Principia*," in his edition, cited above, of the *Principia*, 278. Iliffe suggests that Newton's calculated reticence in the *Principia* may be his own version of the priestly guardianship of esoteric knowledge in the *prisca scientia* and *prisca theologia* traditions that, as we shall see, Newton was not only keenly interested in but indeed considered himself a member of (*Priest of Nature*, 200, 218).

[22] Newton, *Principia*, 382; my emphasis.

[23] Levitin, *Kingdom*.

[24] Levitin, *Kingdom*, 168.

[25] Levitin, *Kingdom*, 169.

[26] Levitin, *Kingdom*, 171–74.

philosophers—men like Xenophanes, Melissus, and Parmenides, but also theogonic poets like Hesiod—were monist materialists who thought the world emerged from one material first principle, certainly very different from the Judaeo-Christian deity."[27]

G. J. Vossius in his magisterial *De theologia gentili, et physiologia Christiana sive de origine ac progressu idololatriae* (1641, rev. 1668), Newton's dog-eared copy of which survives,[28] further undermined the case for the compatibility of Christian and pagan philosophy and theology by demonstrating that ancient pagan theology was characterized by what we would anachronistically call animism, pantheism, or vitalism.[29] If Patristic apologists saw pagan philosophers and theologians arriving at some knowledge of God from observing the natural world, Vossius countered that they did so at the expense of identifying God with nature rather than seeing God as transcendent. The logic of paganism, depending on natural reason and ignorant of divine revelation, led to "an animism that concealed an imperfect monotheism."[30] It was left to Pierre Gassendi to put the nail in the coffin of Christian-pagan philosophico-theological compatibility, by emphasizing the animism of pagan thought as antithetical to Christian ideas of transcendence and turning instead to the mechanist atomism of Epicurus, which he modified by stipulating that the Christian God first set the atoms in motion. Gassendi was not specific on how the transcendent God animated or moved the atoms, beyond the suggestion that God invested the atoms with the power of motion by superaddition, a conception that, we shall see, informs Newton's speculations on active principles. Newton likewise adamantly affirmed a transcendent God's action in nature without feeling obliged to conclude precisely how God invested matter with the power of motion.

Robert Pasnau has documented the revolt against dualist metaphysics central to Levitin's picture of the emergence of early modern natural philosophy and theology. Pasnau tracks the replacement of Aristotelian hylomorphic understanding of existents as composed of form and matter by a corpuscularian model. Pasnau differentiates integral parts (e.g., a plant's stem, leaves, and flower) from metaphysical parts (a plant's form and matter). Integral parts "are the parts of a body that are themselves bodies. Metaphysical parts, in contrast, are not bodies but are instead the ingredients of bodies."[31] One can separate the integral parts of a body, but one cannot separate the metaphysical parts. These parts can be, but need not be, atoms; any model of body divisible into parts would work, including Descartes' non-atomist, vacuumless *plenum*. The replacement of metaphysical parts by integral parts, which in Pasnau's terms is the replacement of a dualist metaphysics by corpuscularianism, addressed a problem that had bedeviled

[27] Levitin, *Kingdom*, 173.
[28] Levitin, *Kingdom*, 581, n. 13.
[29] Levitin, *Kingdom*, 180.
[30] Levitin, *Kingdom*, 183.
[31] Robert Pasnau, *Metaphysical Themes: 1274–1671* (Oxford: Oxford University Press, 2011), 9.

ancient and scholastic discussions of Aristotelian prime matter, which is by definition formless and thus quality-less. Philosophers, as Pasnau outlines, debated whether attributing any quality at all, including extension, to prime matter, violated its definition as without form. Corpuscularians' alternative conception of prime matter retained the Aristotelian and scholastic conception of prime matter as being a substratum that is conserved through all change, but now that substratum is conceived as body extended and impenetrable as opposed to a paradoxical and unimaginable matter, or *prima materia*, utterly without qualities, including the most basic qualities of extension and mass.[32] The debate has implications, as we will see in Chapter 1, for understanding the chaos of Milton's *Paradise Lost*, which, I will argue, contains a corpuscularian rather than a metaphysical prime matter. The transformation from metaphysical (form-matter) to corpuscularian (parts-whole) conceptions of existents carried implications also for theology, including for prior conceptions of the soul as the immaterial form of the organic body and for descriptions of the Trinity articulated through metaphysical categories such as essence and substance. If form is becoming obsolete, a new understanding of the soul would be needed, and if metaphysical terms have become suspect, centuries of metaphysical arguments about the Trinity would need to be revisited.

With its challenge to belief in an immaterial soul as the form of the living body, corpuscularianism brought with it the question of what accounted for life. Raphaële Andrault has argued that it was only "within the polemical context of the refutation of Cartesian dualism in the late 17th century that life and vitality have become scientific and philosophical objects in themselves." In medical texts of the period, on the other hand, "life itself is rarely conceived as an object of scientific examination or explanation, but only as a vulgar notion reducible to some more fundamental causes. Life is the *explanandum*, not the *explanans*: an effect, and not a cause or principle."[33] Cartesian dualism, with its denial of the vegetative and sensitive souls, its mechanization of all nature outside of *res cogitans*, and thus outside an immaterial God and the immaterial minds of rational creatures, made the questions of the border between the living and the non-living and of what accounts for the life or apparent life of non-human animals and vegetation inescapable and urgent. Milton's and Newton's responses to Descartes' dualism would resist what Hans Jonas, in his landmark *The Phenomenon of Life: Toward Philosophical Biology*, has described as the "last and lethal metamorphosis" of Cartesian dualism, which was "shortly to disintegrate into the equally barren alternatives of idealism and materialism." Jonas argues that the seventeenth

[32] Pasnau, *Metaphysical Themes*, 18.

[33] Raphaële Andrault, "What is Life? A Comparative Study of Ralph Cudworth and Nehemiah Grew," in *The Life Sciences in Early Modern Philosophy*, ed. Ohad Nachtomy and Justin E. H. Smith (Oxford University Press, 2014), 33.

century witnessed the death throes of an ancient vitalist conception of the world. In that superseded conception, the "intelligible" is derived from the "intelligent," a creating reason that orders the world by means of an immanent soul as cause of both motion and rational design.[34] In the wake of Descartes's segregation of mind from body, of *res cogitans* from *res extensa*, with body or *res extensa* governed by mechanism rather than by ordering forms, the natural world became explicable in terms of the collisions of unreasoning bodies. The result, reversing the older order, was that in natural philosophy the least intelligent, units of passive matter, became the most intelligible.

A natural world explicable by the operation of mechanical laws on passive matter is, Jonas argues, delivered to contingency and necessity: contingency, because the state of the world depends on the inherently aimless concatenation of material collisions; necessity, because the laws of motion acting on bodies have led inexorably to the present state of things. When Descartes in *The Discourse on the Method* and *The World* (*Le Monde*) proposed that given the laws he has outlined the world would have emerged in its present form from a chaos without the intervention of a creating God, he is careful, wary of theological objections, to label the proposal as a counterfactual hypothesis and to grant that we know that God created the world.[35] Descartes' dualism led to a segregation of mind from natural philosophy. The result was a splintering, with idealism and materialism as what Jonas calls the "fragmentary, residual products" of this dualism.[36]

Jonas argues that the new mechanist science was unequal to the phenomenon of life. Just as death was the anomaly in a vitalist world, life is an anomaly in a world understood mechanistically. He points to a crucial difference between living and non-living forms ("forms" now understood not in Aristotle's metaphysical terms but in corpuscularian terms as material organizations). The form of a rock, for example, is the result of geological mechanics. Movements of particles caused by prior motions have led over time to the depositing, concreting, and hardening of particles into sedimentary rock. The form is contingent, arising from a particular collection of particles brought together fortuitously by the laws of motion. In non-living things, "form is no more than a changing composite state, an accident, of enduring matter" (80). In living things, on the other hand, form is the continuing organization, and the matter making up any organism is the transient accident. As Jonas writes, a living body's "material contents in their succession are phases of transit for the self-continuation of the form." The result, for Jonas, is the dialectic of necessity and freedom in living things: the necessity

[34] Hans Jonas, *The Phenomenon of Life: Toward a Philosophical Biology* (Chicago: University of Chicago Press, 1966), 70.

[35] René Descartes, *Discourse on the Method* V, *The World*, Chapter 6, and *Principles of Philosophy* III.45; see my *Milton among the Philosophers: Poetry and Materialism in Seventeenth-Century England* (Ithaca: Cornell University Press, 1991), 28.

[36] Jonas, *Phenomenon of Life*, 23.

of access to matter to replenish itself (through consumption) and the freedom from any particular and temporary material makeup and thus the "freedom...to change its matter" (83). Freedom at its nascent and most primitive level is rooted in metabolism.

To metabolize the organism must be open to, able to sense, and empowered to appropriate material from its surroundings. This openness comprises, in Jonas' words, "the quality of felt selfhood, however faint its voice.... Whether we call this inwardness feeling, sensitivity and response to stimulus, appetition, or nisus—in some (even if infinitesimal) degree of 'awareness' it harbors the supreme concern of organism with its own being and continuation of being" (84). This degree of inwardness, of faintly expressed selfhood, fits well, I will argue, with the matter of Milton's universe, with its "Suns" and "attendant Moons.../ Communicating Male and Female Light, / Which two great Sexes animate the World" (8.148–51), its self-opening gates of heaven, its gestating Earth, and its heavenly matter rejecting hellish matter. For Jonas, this rudimentary freedom, one might say at this level "voluntariness," expands as one moves up the scale of being.

For Jonas, freedom is not merely characteristic of but entailed by life. Although distinguishing (unlike Milton in my view) non-living from living nature, Jonas draws the strong link between life and the possibility of freedom that I have claimed to be the impetus behind Milton's animist materialism, and he helps us understand the kind of rudimentary freedom characterizing the sub-rational world of Milton's creation. In turn, Milton's work and other versions of early modern vitalism, including Newton's, serve as resistance, ill-fated and in Milton's case arguably lacking in rigor, to what Jonas has diagnosed as the pathology of Cartesian dualism.

Jonas observes that "The organic even in its lowest forms prefigures mind, and...mind even in its highest reaches remains part of the organic. The latter half of the contention, but not the former, is in tune with modern belief; the former, but not the latter, was in tune with ancient belief" (1). Milton and Newton enter at the inflection point between the ancient and modern conceptions of the relation between mind and the organic, and in larger terms between mind and matter. It has been my contention that for Milton life characterizes all matter, and that he does not draw the line that Jonas draws between living organic matter and non-living inorganic matter. If this is the case, then for Milton the "felt voice" of "faint selfhood" extends through the scale of being. Newton, I will argue, arrives at a similar though not identical position: all matter is either alive or potentially alive, ready to be kindled into life by the agency of matter already living.

The questions of where life resided had theological implications for Milton, Newton, and their contemporaries. If the principle of life was not an immaterial soul created and infused by an immaterial God but rather immanent in matter itself, they would need to close a door that seemed to open to atheism. A material principle of life could make God seem less necessary, and a merely material universe could seem to lack a connection to an immaterial God. One might say that this was the macrocosmic version of a problem at the heart of Cartesian dualism itself, one Descartes attempted to solve by locating the mysterious interaction of *res cogitans* and *res extensa* in the pineal gland.

Unlike dualists who viewed incorporeal substance as essential to beliefs in reasoning souls, in the existence of God, and in that God's relation to the universe, Milton and Newton found in vital matter created by God the substance of both soul and world. Developments in experimental natural philosophy, most notably in iatrochemistry (or medical alchemy), developments in which Milton and Newton both had not merely interests but also attachment to networks of the experimentalists themselves, had made it possible to contemplate the life of matter itself. The idea proved congenial to both, if for different reasons. Milton, as I've argued in the past, was motivated by the need to find a home for free will, to which he was fiercely committed, at a time when purely mechanist theories of the universe threatened that belief, most urgently in England in the mechanist materialism of Thomas Hobbes. For his part, Newton bridled at the insistence of dualists, including Descartes and the Cambridge Platonists, that body by itself in the absence of incorporeal spirit must be inert, as an affront to the omnipotence of God. Moreover, the manuscript record demonstrates that he was hopeful of discovering material active principles accounting for movement, a pursuit he scrupulously bracketed out of the rational mechanics of the *Principia*. And in their theological writings, both reflected the anti-metaphysical movement, described by Pasnau and Levitin, in their attacks on the corruption of Christian dogma resulting from the importation of Greek metaphysics, a patristic development that issued in the doctrine of the Trinity. Shared interests and ideas, however, did not preclude important differences, which are in some cases driven by Milton's and Newton's diverging theological imperatives.

Free Will and Omnipotence

In tracing the convergences of Milton's and Newton's thought in the following chapters, I will argue that their guiding commitments to free will and divine omnipotence, respectively, inform significant differences on issues on which they otherwise agree. Just as Milton returned again and again to defenses of the freedom of the will, Newton's work on his mathematics of forces in the *Principia*, his study of the nature of matter and material transformations in his alchemical work, and his exegesis of biblical prophecy were all marshaled to demonstrate the omnipotence of God.

While Milton and Newton agree with each other and with Christians generally that God is omnipotent, they differ on the effective scope of the divine will. While Milton affirmed God's omnipotence, he was equally committed to the principle that the law of non-contradiction applied to God as to everything else. One implication for Milton was rejection of *creatio ex nihilo*. The Hebrew word "to create," Milton observed, always means to make *from something*, thus rendering creation from nothing a contradiction.[37] Milton's solution was to identify God as material cause of both chaos and creation. In a very real sense, as John Rumrich has established, God creates the universe out of his own "body" or matter.[38] Disagreeing with Milton and agreeing with virtually all Christians, Newton, undeterred by this possibility of contradiction, accepted that God created *ex nihilo*. Matt Goldish maintains that Newton rejected the position of Francis Mercury van Helmont and of Gottfried Leibniz, shared by Milton, that prime matter emanates from God, in part because it jars with Newton's strong voluntarism, to which creation *ex nihilo* is well-suited.[39]

Milton also argued that one omnipotent action could not contradict another. Thus God, having granted rational creatures free will, as Milton insists that God has, could not then determine a rational creature's choices, even tangentially. In *De Doctrina Christiana*, Milton counters those who argue that a creature's significant freedom of the will would conflict with God's immutability, inasmuch as a free, undetermined choice would have the power to change the course of events foreknown and foreordained by God. He observes that God foreknows the free choices of rational creatures, which to God are certain owing to his omniscience though not determined, because the creature's free choices are logically prior to God's foreknowledge of them ("For neither does anyone believe because God had prescience of it—rather, God had prescience of it precisely because someone was going to believe" [*DDC* I.4, *OCW* 8.1:89]). Milton argues that

> the divine intention is said to stand firm against all human strength and intention, but not against the freedom of the will in things where God himself had made man his own master, and had decreed to do thus from eternity—for otherwise, one of God's decrees would conflict with another, with the very result which you so object to in other people, that God becomes mutable while you make inevitable the things which he had ordered to be left to [human] judgement. (*DDC* I.3, *OCW* 8.1:59–61)

[37] Milton, *De Doctrina Christiana* I.7, in *The Complete Works of John Milton*, Vol. 8, Parts 1 and 2, ed. John K. Hale and J. Donald Cullington (Oxford: Oxford University Press, 2012), 8.1:287–89; further quotations of Milton's theological treatises are taken from this edition and cited parenthetically in the text as *DDC* and *OCW*.

[38] John Rumrich, "Milton's God and the Matter of Chaos," *Publications of the Modern Language Association* 110 (1995): 1035–46.

[39] Matt Goldish, "Newton's 'Of the Church: Its Contents and Implications,'" in *Newton and Religion: Context, Nature, and Influence*, ed. James E. Force and Richard H. Popkin (Dordrecht: Kluwer Academic Publishers, 1999), 163. Goldish notes that Newton did in his earlier *De gravitione et aequipondio fluidorum*, a text I will examine in Chapters 1 and 2, entertain emanationist theory.

For Milton divine immutability is threatened not by the potential for human choices to determine items of divine knowledge but rather by the specter of God's determining the choices of rational creatures to whom he has given free will.

For Milton, freedom in some form can be found at every link in the chain of being. While a rationalist in theology, Milton explicitly opposes the argument by the rationalist Cambridge Platonist theologians that God had to create on the principle that creation, being an unambiguous good, had to be chosen by a benevolent God.[40] Milton's God, as the Father proclaims in *Paradise Lost*, has the option to "put not forth my goodness, which is free / To act or not" (7.171–72). Milton writes in his theological treatise, "Indeed I do not even concede any necessity in God to act, but only that he is necessarily God. Scripture itself witnesses that his decreeing, and likewise whatever action he takes, are absolutely free" (*DDC* I.3; *OCW* 8.1:59). "Absolute" freedom is centrally important to Milton. When it comes to rational creatures, he is adamant that no shred of necessity resulting from divine omnipotence and omniscience can determine rational creatures' choices. Satan's tempting words to Eve in *Paradise Lost*, "Not just, not God" (9.701), though in a sense different from the one meant by the speaker, are true for Milton. Milton's God's insistence in *Paradise Lost* that no hint of fate or foreknowledge causes the Fall—the Father will not "touch with lightest moment of impulse / His free will, to her own inclining left / In even scale" (3.45–47)—echoes the Milton of *De Doctrina Christiana*, who insists that "all necessity must be removed from our freedom, nor even must that shadowy and external necessity based on immutability or prescience be admitted to the discussion" (*DDC* I.3; *OCW* 8.1:61). The concern for Milton explicitly is divine justice. His God cannot punish rational creatures for sins that they have not chosen freely. There are no rational creatures created to be "vessels of wrath." And if freedom goes all the way up to God in Milton's mind, it also at least in rudimentary form goes all the way down to living matter.

Newton has little to say in his theological and exegetical manuscripts on the freedom of the will. Rob Iliffe finds evidence for Newton's endorsement of the freedom of the will in our ability to move our bodies by a mere act of the will, but this demonstration of the power of the will to initiate bodily motion does not preclude the possibility that the will might be determined to choose this or that motion or action; determinists like Hobbes did not deny that we could move our bodies by acts of will.[41] Scott Mandelbrote maintains that Newton held an Arminian position on free will, a conclusion he draws apparently from Newton's

[40] Stephen M. Fallon, "'To Act or Not': Milton's Conception of Divine Freedom," *Journal of the History of Ideas* 49, no. 3 (1988): 425–49.

[41] Iliffe, *Priest of Nature*, 103–5, 120. Iliffe draws his examples from the manuscript "De Gravitatione et Æquipondio Fluidorum" and the 1706 Queries to the *Opticks*.

emphasis on obedience to the commandments of the Noachide law.[42] But predestinarian, anti-Arminian Calvinists also emphasized obedience to commandments, and Calvin could speak of the freedom of individuals lacking grace to choose the sins that they inevitably choose. Goldish is closer to the mark in ascribing to Newton a "vague predestinarianism."[43] In his manuscript "Theological Notebook" (ca. 1684–90), Newton writes,

> Is there unrighteousness with God? God forbid. For he saith to Moses, I will have mercy on whome I will have mercy, & I will have compassion on whom I will have compassion. So then it is not of him that willeth, or of him that runneth, but of God that sheweth mercy [Romans 9:16]. For the scripture saith unto Pharoh Even for this same purpose have I raised thee up that I might shew my power in thee... Therefore he hath mercy on whom he will have mercy, & whom he will he hardneth.[44]

In this passage, Newton articulates an anti-Arminian endorsement of particular predestination, as opposed to the Arminian (and Miltonic) position of the general predestination to salvation of all who freely choose to believe. The difference can be seen in Milton's own discussion of Romans 9:16:

> when God had decided to restore humankind, he also indubitably decreed (for what was fairer?) to restore lost freedom to the will, in some measure at least. So to those whom, in accord with that choice,... he saw in whatever way wishing or else running (it is probable that they are here called 'ordained'), to them he gave the ability to wish and run more fully, that is, to believe. (*DDC* I.4, *OCW* 8.1:95)

Newton follows Calvin in going on to cite the verses closely following ("Nay but, O man, who art thou that repliest against God? Shall the thing formed say to him that formed it, Why hast thou made me thus?" [Romans 9:20]) and to rebuke those who would question the justice of God in condemning to suffering those not free to choose rightly. Milton, on the other hand, ignores the verse on the potter and the clay and characteristically emphasizes the justice of God that would be called into question by the position taken by Calvin and Newton.[45]

[42] Scott Mandelbrote, "'A Duty of the Greatest Moment': Isaac Newton and the Writing of Biblical Criticism," *The British Journal for the History of Science* 26, no. 3 (1993): 292.
[43] Goldish, "Newton's 'Of the Church,'" 161.
[44] Newton, "Theological Notebook, Part One," Keynes Ms. 2, 31r; https://www.newtonproject.ox.ac.uk/view/texts/normalized/THEM00180, accessed January 17, 2024.
[45] While Milton will conclude that God is just, the great argument of *Paradise Lost* approaches the question of divine justice as urgent and real rather than as blasphemously presumptuous.

In the same manuscript, however, Newton does offer a more qualified perspective on human freedom: "{A sinn}er does evil not because he cannot do all things that he will but becaus he will not do all things that he can ¶ His wil against evil is determined not by any blind impuls of his nature but by free election. ffor evil is folly & is always avoided by a perfect understanding."[46] Newton here suggests that a perfect understanding can prevent one from sinning, contrary to Sir Philip Sidney's argument that our "erected wit maketh us know what perfection is, and yet our infected will keepeth us from reaching unto it."[47] The implication may be that we are free to choose good or evil, that none of us is a "vessel of wrath" made "unto dishonor" as Newton writes in the same Notebook. But it may also be the case that Newton, like Calvin, believes that those from whom grace is withheld (and thus who cannot be saved) can be said to choose their faithlessness.

Across his remarkably varied intellectual explorations and disciplinary interventions, Newton focuses relentlessly on God's omnipotence. The *locus classicus* for Newton's emphasis on God's omnipotence is the 1713 General Scholium:

> He rules all things, not as the world soul but as the lord of all. And because of his dominion he is wont to be called Lord God *Παντοκράτωρ*/Pantokrator. For "god" is a relative word and has reference to servants, and godhood is the lordship of God, not over his own body as is supposed by those for whom God is the world soul, but over servants. The supreme God is an eternal, infinite, absolutely perfect being; but a being, however perfect, without dominion is not the Lord God. For we do say my God, your God, the God of Israel, the God of Gods, and Lord of Lords, but we do not say my eternal one, your eternal one, the eternal one of Israel, the eternal one of the gods; we do not say my Infinite one, or my perfect one. These designations do not have reference to servants. The word "god" is used far and wide to mean "lord," but every lord is not a god. The lordship of a spiritual being constitutes a god, a true lordship constitutes a true god, a supreme lordship a supreme god, an imaginary lordship an imaginary God.[48]

Newton stipulates a being eternal, infinite, and perfect, but these essential attributes are not what constitute God as God. Instead, Newton's focus is on God as he relates to his creation, and his relation to that creation is defined by his power and dominion.

As the following chapters will illustrate, Milton's emphasis on the freedom of the will and Newton's on divine omnipotence will mark their overlapping

[46] Newton, "Theological Notebook (Part 2)," Keynes Ms. 2, A^{v}; https://www.newtonproject.ox.ac.uk/view/texts/normalized/THEM00181, accessed August 6, 2025. The part in brackets is the editor's reconstruction of missing text.

[47] Sir Philip Sidney, "A Defence of Poetry," in *The Miscellaneous Prose of Sir Philip* Sidney, ed. Katherine Duncan-Jones and Jan van Dorsten (Oxford: Oxford University Press, 1973), 79.

[48] Newton, *Principia*, 940–41.

systems of thought. For Milton the vitality of matter will support his defense of freedom against the inroads of mechanistic determinism. In Newton's case, vital matter will carry the continuing creative efficacy of the omnipotent God, who seeds the world with what Newton calls his "protoplasts." The prophetic Milton focuses on filling in gaps in Scripture and on discerning the spiritual significance of Scripture as a guide to the moral and liturgical choices of individuals; Newton will focus on the demonstration of hitherto unrecognized evidences of God's determining power in the historical record. The diverging theological emphases of the two will inform how they understand the Son's relation to the Father and to human beings and will influence which aspects of Socinian teaching they will approach or adopt.

Networks

In exploring the contexts of Milton's intellectual world, I will look to figures other than Newton. Milton's animist materialism participates in an efflorescence of vitalism in seventeenth-century England, and illuminating parallels to Milton's thought will be found in Francis Bacon; in Lady Anne Conway, whose work I have addressed previously[49]; in the fiction and philosophy of Margaret Cavendish, the Duchess of Newcastle; and in Francis Glisson, all of whom will appear below. The antitrinitarian Milton will find possible allies in John Locke and in Socinian theologians. My main focus, however, will be Newton, for two reasons. The first and most important is the extent of the parallels in their self-understandings as prophets, in their vitalist speculations, and in their Christological Arianism and similar gestures toward Socinianism. The second reason is instrumental. It can be tempting to consign Milton's vitalist world to poetic fancy, a groundless invention by a poet dabbling in natural philosophy. The antitrinitarianism of *De Doctrina Christiana* has been wrestled into orthodoxy, challenged as the work of someone else than the great Christian poet, or seen as a crochet not infecting *Paradise Lost.* By presenting the related theological and natural philosophical views of Milton as uncannily similar to the thinking of one of the greatest minds of seventeenth-century England, I hope to show that Milton is not a dilettante trading in dead ideas but one engaged in and contributing to intellectual developments alive and influential in the early modern world. My hope is that it will be harder to condescend to his thought while praising his poetry if that thought resembles Newton's.

While I will demonstrate the broad and, in some areas, deep resemblance between the thought of Milton and Newton, I will not argue for direct influence. Neither mentions the other in his writing. The first work of Newton's to appear in

[49] Fallon, *Milton among the Philosophers*, Chapter 4, "Milton and Anne Conway," 111–36.

print, and the only one to be published in Milton's lifetime, was his "New Theory about Light and Colors," published by the Royal Society in February of 1672, less than three years before Milton's death, and two decades after he had gone blind. Even though Milton may have kept up with mathematics in his earlier years, it is hard to imagine that he could have absorbed much of this technical work, with its elaborate diagrams, had it been read to him. Newton did own a copy of Milton's poems, but in a two-volume 1720 quarto edition, which is far too late to have influenced his thinking on shared topics of interest.[50] He could have read Milton earlier, in a lost or borrowed volume, but Newton's reported reliance on the characterization of poetry of Isaac Barrow, his teacher and predecessor as Lucasian Professor of Mathematics, that "poetry was a kind of ingenious nonsense," is not encouraging to anyone looking for evidence of the poet's influence on the natural philosopher.[51]

There is no more reason to think that they would have met in London or Cambridge. Milton left the university in 1632, ten years before Newton's birth, and he never returned. While Newton visited London three times during the last six years of Milton's life,[52] it is unlikely that the obscure young natural philosopher would have looked up the aging poet. On the other hand, it is equally unlikely that Newton was oblivious to the appearance of a celebrated epic in which the actions of the creating Christian God are played out on a cosmic stage.

It is in any event probable that Newton was aware of some of Milton's ideas. He would certainly have known the notorious and celebrated name of the defender of the execution of Charles I and of the poet of *Paradise Lost*. And they did have friends and acquaintances in common. Katherine Jones, Viscountess Ranelagh, the sister of Newton's correspondent Robert Boyle, was a particularly close friend of Milton[53]; she arranged for Milton to tutor her son, Richard Jones. She had scientific interests of her own, and it is likely that the Invisible College, the pre-Restoration predecessor of the Royal Society, met at least some of the time in her home. It is possible that she mentioned something of Milton's ideas to Newton. The same can be said of Milton's friend Henry Oldenburg, another member of the Milton-Ranelagh circle, one of the founders of the Royal Society and its first Secretary, who played a key role in persuading Newton to submit his papers on optics to the Society and to allow them to be published in its journal.[54]

[50] John Harrison, *The Library of Isaac Newton* (Cambridge: Cambridge University Press, 1978), 71, 192.

[51] Westfall, *Never at Rest*, 581. Harrison's *Library*, however, contains evidence of Newton's possession of classic English poets apparently unknown to Westfall.

[52] Newton spent a month in London in August–September of 1668, under two weeks in November–December 1669, and a week in August–September 1674, two months before Milton's death (Westfall, *Never at Rest*, 189, 196, 256).

[53] Lady Ranelagh petitioned for mercy on Milton. See forthcoming work by Alison Chapman.

[54] Marie Boas Hall, "Oldenburg, Henry [Heinrich] (c. 1619–1677)," *Oxford Dictionary of National Biography*; odnb 9780198614128, accessed August 25, 2025.

Newton and Oldenburg were frequent correspondents during the last years of Milton's life, with two dozen letters from Newton to Oldenburg and fourteen from Oldenburg to Newton dated between January 2, 1671/72 and September 14, 1673.[55] Still another member of the Ranelagh Circle, and a "familiar acquaintance" of Milton, Samuel Hartlib, at whose prompting Milton wrote *Of Education*, was the center of a circle of alchemists, from whom Newton may have obtained alchemical manuscripts.[56] They are connected then, at least indirectly, by webs of friendship and professional and personal acquaintance, which raises the possibility, at least, that Newton may have known something of Milton's theological and natural-philosophical ideas.

Newton's Manuscripts

The natural philosophical and Christological ideas of Milton that I will focus on can be found mainly in two works easily accessible to and familiar to students of Milton, the epic *Paradise Lost* and Latin treatise *De Doctrina Christiana*. The authorship of the epic is clear. Though the authorship of the treatise has been challenged, for reasons that I examine in Chapter 4 below, the challenges can be traced, I will suggest, to reticence to accept that the greatest Christian epic was written by someone who from an orthodox perspective was a heretic. For reasons that I have addressed elsewhere, I accept Milton's authorship of the treatise as firmly established.[57]

The Newton whose ideas resonate with Milton's, the insistently antitrinitarian and philosophically speculative Newton, is found in the manuscripts. If the Newton of the *Principia Mathematica* famously does not feign hypotheses, the Newton of the manuscripts speculates and hypothesizes. In the religious manuscripts he advances an Arian understanding and flirts with a different Socinian understanding of the Son of God, ideas that he carefully concealed from the public.

Newton's manuscript legacy has become available through two related efforts. The Newton Project, founded at the University of Sussex in 1998 under the editorship of two leading Newton scholars, Iliffe and Mandelbrote, was launched with the goal of producing a comprehensive online edition of Newton's "non-scientific" papers. Having moved with Iliffe to the University of Oxford, the project's goal has expanded, the goal now being "a comprehensive edition of all

[55] See the letters in *The Correspondence of Isaac Newton, Volume 1: 1661–1675*, ed. H. W. Turnbull (Cambridge: Cambridge University Press, 1959).

[56] For Hartlib's friendship with Milton, see Lewalski, *Life of John Milton*, 238; for the Hartlib-Newton nexus, see Westfall, *Never at Rest*, 288.

[57] Fallon, "Milton's Arminianism and the Authorship of *De Doctrina Christiana*," *Texas Studies in Literature and Language* 41 (1999): 103–27. And see below, Chapter 4, n. 18.

of Newton's printed and unpublished writings."[58] Newton's religious, theological, and historical manuscripts, mainly in Latin and English and gathered in The Newton Project, contain more than three million words. Partly accounting for this massive word count is Newton's obsessive recopying and redrafting of his major manuscript works, notably the "Treatise on Revelation" (1670s–80s) and the "Theologiæ Gentilis Origines Philosophicæ" (1684–90), as well as drafts for parts of his two posthumously published books, *The Chronology of Ancient Kingdoms Amended* (London, 1728) and the *Observations on Prophecies of Daniel, and the Apocalypse of St. John* (London, 1733).[59] This copying and redrafting testifies to Newton's obsessive interest in theological and religious questions. The manuscripts range in date from the 1660s to well into the eighteenth century. As the third and fourth chapters will show, parallels with Milton's thought appear across this range of composition. Interested readers can find full texts of these manuscripts in the online Newton Project.

Newton became interested in alchemy in his twenties, around 1669,[60] and he was an active student and practitioner of alchemy at least through the 1690s. Newton's alchemical manuscripts, which are conveniently linked to the Newton Project, have been gathered under the editorship of William R. Newman at Indiana University in the second project, The Chymistry of Isaac Newton ("chymistry" is a term often used in the seventeenth century for alchemy). This project comprises "a rich and diverse set of document types, including laboratory notebooks, indices of alchemical substances, and Newton's transcriptions from other sources."[61] The manuscripts address both chrysopoeia, the theory and practice of the transformation of metals, and iatrochemistry, the application of chemical principles to medicine. The alchemical manuscripts contain more than 540,000 words. While both manuscript collections contain Newton's transcriptions of and notes on texts by others, the proportion of such texts is vastly higher in the alchemical archive, but his alchemical papers do include laboratory notes on his own experiments, in a laboratory he built himself.[62] The best window into his alchemical speculations can be found in two manuscripts, which I will discuss in Chapter 2, that appear on reverse side of the same folio sheets, "Of natures obvious processes in vegetation" and "Humores minerales," which The Chymistry

[58] "About the Newton Project," https://www.newtonproject.ox.ac.uk/about-us/newton-project, accessed May 16, 2025.

[59] Several important shorter manuscripts on religion were also collected and published by Herbert McLachlan in *Sir Isaac Newton's Theological Manuscripts* (Liverpool: At the University Press, 1950).

[60] Westfall, *Never at Rest*, 290.

[61] https://webapp1.dlib.indiana.edu/newton/.

[62] See, e.g., an extensive, twenty thousand word set of notes on his own experiments in the winter of 1678–79, Cambridge University Library Ms. Add. 3973. The leading scholars of Newton's alchemical experiments, both of whom have reproduced some of Newton's own alchemical experiments, are Lawrence Principe, *The Secrets of Alchemy* (Chicago: University of Chicago Press, 2013) and William R. Newman, *Newton the Alchemist: Science, Enigma, and the Quest for Nature's "Secret Fire"* (Princeton: Princeton University Press, 2019).

of Isaac Newton dates to the first half of the 1670s.[63] Other, earlier manuscripts were undoubtedly lost in a fire in Newton's laboratory in the winter of 1678–79 that destroyed many of his alchemical papers.

Prisca Theologia and *Prisca Scientia*

In his varied intellectual pursuits, including his mathematical physics as well as his alchemy and his theology, Newton saw himself as a radical in the literal sense, as one returning to ancient roots. Milton, along with most Protestants and particularly Puritans, was intent on restoring the practices and beliefs of primitive Christianity. Newton looked further into the past. Newton was committed, like others in the seventeenth century and in particular like his older friend the Cambridge Platonist Henry More,[64] to recovering the thought of the most ancient philosophy and religions, the *prisca scientia* and the *prisca theologia* that made up the *prisca sapientia*.[65] In this, his thought was more radical than Milton's. Eagerness to recover and reenshrine ancient wisdom and religion fuels his repetitive and even obsessive writing and rewriting of ancient secular and religious histories.

Newton viewed his varied intellectual interests, in natural philosophy, biblical hermeneutics, theology, and history, as part of a unified intellectual program recovering the most ancient wisdom.[66] This program, Newton believed, descended from and was focused on restoring the religion of the second founders of the human race, Noah and his sons, a religion adopted by the ancient Egyptians. He believed that the ancients had understood and demonstrated the heliocentric arrangement of the solar system and the inverse square law of gravitation,[67] as well as God's substantial omnipresence in the universe.[68] In several so-called "classical

[63] Dibner Ms. 1031 B, Smithsonian Institution, Washington, DC; https://webapp1.dlib.indiana.edu/newton/mss/dipl/ALCH00081. This book's frontispiece reproduces the first page of "Of natures obvious laws and processes in vegetation" as an example of Newton's generally legible hand.

[64] Westfall, *Never at Rest*, 55–56. More in his will left Newton a funeral ring ("The Will of Henry More," October 2, 1687, reprinted *The Conway Letters: The Correspondence of Anne, Viscountess Conway, Henry More, and Their Friends*, ed. Marjorie Hope Nicolson, 1930; rev. ed., ed. Sarah Hutton [Oxford: Clarendon, 1992], 482).

[65] Levitin cautions against the belief that seventeenth-century thinkers, including Newton, intent on recovering the *prisca sapientia* accepted a full suite of what were in fact different views: the derivation of all philosophy from Moses, Noah's religion as the Ur-religion, Hebrew foreshadowing of trinitarianism, and pagan monotheism. See Dmitri Levitin, *Ancient Wisdom in the Age of the New Science: Histories of Philosophy in England, c. 1640–1700* (Cambridge: Cambridge University Press, 2015), 8–16.

[66] I will argue in this book for important links between Newton's theology and natural philosophy. For arguments against the unity of Newton's thought, see Rob Iliffe, "Abstract Considerations: Disciplines and the Incoherence of Newton's Natural Philosophy," *Studies in the History and Philosophy of Science* 35 (2004): 427–54.

[67] J. E. McGuire and P. M. Rattansi, "Newton and the 'Pipes of Pan,'" *Notes and Records of the Royal Society of London* 21, no. 2 (1966): 108–43.

[68] Levitin, *Ancient Wisdom*, 436–37.

scholia" to Propositions IV through IX of the Third Book of the *Principia*, unpublished during his lifetime, Newton laid out the case for the ancients' knowledge of the truth of key Newtonian "principles that matter is atomic in structure and moves by gravity through void space; that gravitational force acts universally; that gravity diminishes in the ratio of the inverse square of the distance between bodies; and that the true cause of gravity is the direct action of God."[69] According to Newton, Egyptian natural philosophers and the pre-Socratic philosophers Thales and Democritus had pioneered the knowledge that he was in the process of reviving. Newton agreed with More that Democritus articulated a true natural philosophy of atomism, but he disputed More's charge that Democritus corrupted atomism in an atheistic direction. J. E. McGuire and P. M. Rattansi argue that Newton's discussion of the ancient precedents to his natural philosophy in the *scholia* are integral to his thinking rather than rhetorical flourishes to embellish what he considered his own discoveries.

In Newton's eyes, this true natural philosophy descended to the pre-Socratics from Noah and his followers and then from Egyptian priests. He regarded the Egyptian priests as the keepers and promulgators of the true philosophy, who concealed behind figures and allegories philosophical truths from the vulgar, who were unequal to these mysteries. He was committed to recovering the ancient inseparability of theology, religion, and natural philosophy.

Newton thus was intent on reviving not only the most ancient and accurate natural philosophy but also the most ancient and true religion, an endeavor that would have implications for his radical and sometimes inconsistent recasting of Christian theology. Newton found traces of this ancient religion in the cosmological form of the rites of ancient Egypt and of ancient Rome, the latter established by Numa Pompilius, the legendary second king of Rome, in the eighth century BCE.[70] These rites were centered on temples arranged around a central fire; the twelve elements of the temples figured the sun, the six planets or wandering stars (including the Earth), the four elements, and the quintessence. These temples, or *prytanea*, were conceived as images of the cosmic system, which was itself the one true temple of the creator God, himself imaged by the sun.[71]

Newton was convinced that, over time, the *prisca scientia* of a universe moved by God and the *prisca theologia* of a monotheistic God, whose hand could be discerned in the ordering of the universe and who was to be worshiped in temples that were simulacra of that universe, became corrupt in tandem. The celestial bodies of the cosmic temple figured schematically in worshippers' *prytanea* came

[69] McGuire and Rattansi, "Newton and the 'Pipes of Pan,'" 112. The *scholia* can be found in Gregory Ms. 247 at the Royal Society.

[70] See Richard Westfall, "Isaac Newton's *Theologiae Gentilis Origines Philosophicae*," in *The Secular Mind: Transformation of Faith in Modern Europe*, ed. W. Warren Wagar (New York: Holmes & Meier, 1982), 15–34.

[71] Levitin, *Ancient Wisdom*, 192–93, 436–37.

to be viewed not as aspects of the creation of the one God but as gods themselves, and furthermore they came to be identified with deified rulers and ancestors. Polytheistic idolaters began to worship the celestial bodies as gods and to adore the simulacra of those bodies in *prytanea* as themselves divine. They also began to worship earthly rulers and leaders, for example, Noah and his sons, as translated into stars.[72] The result was a development that horrified Newton and that he viewed his own work as countering, the corruption of ancient monotheism and the fantastic perversions of polytheism.

In Newton's reading, the original, uncorrupt Egyptian and earlier Noachide religions had only two precepts: love of the one God and love of one's neighbor. History in Newton's eyes became a continuing story of the corruption of this primitive religion and the appearance of prophets, among whom he counted himself, pointing humanity back to religion's earlier, pure form. To this conception of a primitive ur-religion can be traced the seed of the most radical element of Newton's religious thought, the demotion of the Son of God and the antitrinitarianism that he would share with Milton. Westfall argues that in his reading of the history of religion, as articulated in a manuscript of the 1680s, the "Theologiae Gentilis Origines Philosophicae," Newton "reduced the coming of Christ from the epochal event of human history to one in a series of attempts to purge idolatry and restore the true worship of God.... [I]t gave central importance to the concept of an original pure religion to which Christian revelation added nothing whatever."[73] While, as we will see in Chapter 4, Newton would endorse the Apostles Creed's belief in "Jesus Christ, His only Son, Our Lord, Who was conceived by the Holy Ghost, born of the Virgin Mary, suffered under Pontius Pilate, was crucified; died, and was buried," Newton would also suggest that one's religious obligation could be fulfilled by adherence to the two commandments to love God and love one's neighbor as oneself. In his manuscript "A short Schem of true Religion," he distinguishes what is fundamental in religion from what is mutable; the former is "the Religion of Adam, Enoch, Noah, Abraham Moses Christ & all the saints & consists of two parts our duty towards God & our duty towards man or piety & righteousness, piety which I will here call Godliness & Humanity."[74] Newton in May 1694 remarked to David Gregory, as summarized by Iliffe, that Christ "restored the religion of Moses, itself a restoration of true elements of worship that were embedded within the Egyptian religion."[75] The manuscript does not single out belief in the divine sonship of Jesus as a defining element of the true religion. In this remarkable manuscript and elsewhere,

[72] See Iliffe, *Priest of Nature*, 213–15.

[73] Richard Westfall, "Newton's Scientific Personality," *Journal of the History of Ideas* 48, no. 4 (1987): 569.

[74] Newton, "A Short Schem of the True Religion," Keynes Ms. 7, f. 1^r; https://www.newtonproject.ox.ac.uk/view/texts/normalized/THEM00007, accessed October 5, 2022.

[75] Iliffe, *Priest of Nature*, 216.

Newton not only entertained an idea of salvation outside the Christian communion but also proposed that monotheistic belief and ethical conduct by themselves were sufficient for salvation. Newton's radically focused insistence here on recognition of the one God as lord of the universe will have implications, as we shall see, for his natural philosophy of gravitating matter, for his task as a prophet, and for his views on the relation of the Son of God to the one God the Father and to rational creatures.

As we turn to the many ways in which the thoughts of Milton and Newton converge, in their advocacy for the life of matter, in their propagation of alchemical understandings of the universe, in their assuming the prophetic mantle, and in challenging Christian orthodoxy on the full divinity of the Son of God, we should keep in mind the two ways in which they stand apart from each other. Milton is the prophet of freedom, the freedom of God, the freedom of rational creatures, and even the freedom of apparently inanimate matter, while Newton is the prophet of divine omnipotence and dominion. And Newton is finally the more radical of the two, in his attempt to restore the ancient philosophy of the pre-Socratics and the ancient religion of Noah.

Milton and Newton: Matter, God, and Modernity. Stephen M. Fallon, Oxford University Press.
 DOI: 10.1093/9780198988939.003.0001

PART I
NATURAL PHILOSOPHY

1
The Life of Matter

In the mid- and late seventeenth century, fundamental questions were contested: what is the divinely created universe made of? What is the relation of body and soul? Which existents are living and which are inert or merely passive? What is the source or cause of life in living beings? In approaching these questions, Milton, Newton, and their contemporaries were intent on squaring their answers with the inspired truths of the Bible. Milton's and Newton's responses resembled each other's in significant and precise ways, and they furthermore reflected the tenor of particularly English experimental philosophy of the early and mid-century. The mechanism characteristic of Descartes' understanding of *res extensa* and of Hobbes' theory of body, which in different ways proposed the passivity of extended stuff, was viewed skeptically by English natural philosophers and experimentalists for reasons that Milton and Newton shared. Like many of their contemporaries, Milton and Newton speculated that life is a property not merely of certain mechanical arrangements of body, but of the basic parcels of matter themselves. For them the universe lives not because of the presence of incorporeal spirit or the mechanistic motions of combinations of matter, but because matter can itself be instinct with life. In this chapter I will explore Milton's and Newton's intimations of the life of matter, and in the next I will turn to the particularly alchemical aspects of their thought.

What Is Life in the Seventeenth-Century England?

As I will be addressing what I will argue is living matter or matter potentially alive in the absence of incorporeal substance in Milton and Newton, I should specify here what I will mean by life. The short version is that for the purposes of my argument the hallmark of life is self-activity. In the seventeenth century as now the definitions of life and of the line separating the living from the non-living were contested. Unlike now, most seventeenth-century metaphysicians and natural philosophers were deeply concerned with the theological implications of how one might answer these questions. One influential line of thinking, championed at great length by Ralph Cudworth in his massive *True Intellectual System of the Universe* (1678), tied life to an incorporeal soul and thence to an incorporeal God. Cudworth's great enemy was hylozoism, the position that material substance, in the absence of incorporeal substance, could move itself, perceive, act, and thus live. This was for Cudworth the royal road to atheism:

> if all Matter, as such, have not only such a *Life, Perception* and *Self-active* Power in it, as whereby it can Form it self to the best advantage, making this a Sun and that an Earth or Planet, and fabricating the Bodies of Animals most Artificially; but also can improve itself into Sense and Self-enjoyment; it may as well be thought able to advance it self higher, into all the Acts of *Reason* and *Vnderstanding* in Men: so that there will be no need either of an Incorporeal Immortal Soul in Men, or a Deity in the Universe.[1]

A universe entirely and exclusively material left no place or role for an entirely immaterial, incorporeal God. For Cudworth, as for his Cambridge colleague Henry More, divinely created incorporeal substance was necessary not merely for the stuff of rational, immortal souls, and not even merely for the vegetable and animal souls of plants and sentient creatures, but for the cause of any and all movement in the world. At the lowest level (Cudworth's "plastic nature" or More's "Spirit of Nature"), this incorporeal substance made possible motion in what would otherwise be an inert, passive, and unmoving mass of spiritless body. To admit into the universe, however narrowly, inherently self-moving matter or matter moved merely mechanically is to step onto a slippery slope at the bottom of which was a denial of soul, immortality, and God.

A less dangerously atheistic alternative for Cudworth was the Christianized atomism of Gassendi, who proposed that God at creation superadded motive power to material. In so investing it, the Christian God did not cede his creative direction of the universe to mere matter, but rather invested atoms with those motions that would issue in his creation. While Cudworth distinguished this modified atomism from what he viewed as the atheistic ancient atomism of Democritus, he remained gravely concerned about any philosophy that did not place incorporeal substance behind all movement, action, and life.

For Cudworth (and for More) extended matter was in itself passive, movable but not self-moving, and dead. Life, on the other hand, is a quality of incorporeal substance. While at its highest reaches in creation, incorporeal substance is responsible for "cogitation" or reason,[2] it is "self-activity" that is the foundational and universal attribute of life: "[I]f there were once no *Motion* at all in the whole world, nor no Life or Self-Active Power in any thing, but all were dead; then is it certain, that there could never possibly arise, any *Motion* or *Mutation* in it to all Eternity. There being no *Sufficient Cause*, to Produce the Same; since nothing can produce *Motion*, but that which hath *Life* or *Self-Activity* in it."[3] Here Cudworth equates life and self-activity ("that which hath *Life* **or** *Self-Activity*"). There are living things that do not think or "cogitate," but everything that lives is self-active,

[1] Ralph Cudworth, *The True Intellectual System of the Universe* (London: Printed for Richard Royston, 1678), 105; Early English Books Online, ProQuest doc. 2264206863, accessed December 5, 2023.

[2] Cudworth, *True Intellectual System*, 27.

[3] Cudworth, *True Intellectual System*, 739.

and by the same token anything self-active is in Cudworth's view alive. He ties life to the self-activity of incorporeal substance, whether in the animal and rational souls of sentient creatures or the incorporeal "plastic nature" responsible for all activity in "whole Terrestrial (or Terraqueous) Globe, by which all Plants and Vegetables . . . may be differently formed, according to their different Seeds, as also Minerals and other Bodies framed, and whatsoever else is above the Power of Fortuitous Mechanism."[4] Cudworth strictly limits the reach of mechanism. A billiard ball struck by another would be moved mechanically, but mechanism could not account for creation, for order, for living organisms, for the production of metals, for any organization of matter, for all of which incorporeal substance is necessary.

In the next century, self-activity was foundational in Immanuel Kant's definition of life as well: "*Life* is the faculty of a *substance* to determine itself to act from an *internal principle*."[5] Elsewhere Kant writes that "Life is the capacity to initiate a state (of oneself or another) from an inner principle."[6] According to Tyke Nunez, for Kant "self-activity will characterize the power of a living being to determine itself, with respect to motion and rest."[7]

Self-activity, or as she terms it "self-motion," is also central to the definition of life for Milton's and Newton's contemporary Margaret Cavendish. A thoroughgoing vitalist who argues that all matter is alive, Cavendish sees self-motion as at the heart of the definition of that living matter: "no motion can be without perception, because every part or particle of nature, as it is self-moving, so it is also self-knowing and perceptive; *for matter, self-motion, knowledge and perception, are all but one thing,* and no more differing nor separable from each other, then body, place, magnitude, colour and figure."[8] Knowledge and perception are themselves instances of the incessant self-motion and thus the life that characterizes every particle of nature. Cavendish can serve as the limit case of seventeenth-century vitalism. Every particle of matter moves itself, perceives, and knows, actively arranging itself in the larger bodies to which it temporarily belongs.[9] I will suggest

[4] Cudworth, *True Intellectual System*, 171.

[5] Immanuel Kant, *Metaphysical Foundations of Natural Science*, trans. and ed. Michael Friedman, *Cambridge Texts in the History of Philosophy* (Cambridge: Cambridge University Press, 2004), 83.

[6] Immanuel Kant, "Notes on Metaphysics," in *Notes and Fragments*, ed. Paul Guyer, trans. Curtis Bowman and Paul Guyer, *Cambridge Edition of the Works of Immanuel Kant* (Cambridge: Cambridge University Press, 2005), 88.

[7] Tyke Nunez, "Kant on Plants: Self-Activity, Representations, and the Analogy with Life," Ann Arbor, MI: Michigan Publishing, University of Michigan Library, *Philosopher's Imprint* 21, no. 11 (2021): 5.

[8] Margaret Cavendish, *Observations upon Experimental Philosophy*, ed. Eileen O'Neill, *Cambridge Texts in the History of Philosophy* (Cambridge: Cambridge University Press, 2001), 113 (I.27); my emphasis.

[9] This is the case even for what Cavendish calls, paradoxically, the "inanimate part of matter": which, "having an innate knowledge and *life of itself*, it is not improbable but it may also have an interior, fixt and innate knowledge of the existency of God" (*Observations* 38 [Argumental Discourse]). For Cavendish on the "inanimate degree of matter" that nevertheless lives, see my "Imagining Animist Materialism in *Paradise Lost* and *The Blazing World*," in *Margaret Cavendish and John* Milton, ed. Lara A. Dodds, Lisa Walters, and Ann Baynes Coiro (Oxford: Oxford University Press, forthcoming).

that Milton imagines a world that comes close to Cavendish's end of the vitalist spectrum, and that Newton contemplated the likelihood that, at the least, all matter could be kindled into life by *material* active principles rather than a separate incorporeal substance.

While there were more maximal definitions of life available in the seventeenth century, including the ability to metabolize, to maintain one's form, and to reproduce, self-activity was either expressly stated or implied in any definition of life. Timothy M. Harrison has recently offered a provocative and maximal new definition of life as conceived by Milton, the felt endeavor of an organism.[10] In my reading of Milton and Newton, I will employ an admittedly more minimal definition, but one dominant in early modern thought—embraced by thinkers as disparate and disagreeing among themselves as Cudworth and More, Cavendish, and Kant—and treat *self-activity* as the criterion of life.

Milton's Vitalism

Milton developed his version of vitalism, which I have labeled animist materialism, in response, as I argued some time ago, to the threat to belief in the freedom of the will posed by competing metaphysical models in the mid-seventeenth century.[11] Exploration of Newton's writings on the nature of body and spirit, and particularly his alchemical works, has led me to revisit my thesis, correcting overstatement of Milton's idiosyncrasy and reversing my earlier argument for Milton's dismissal of alchemy. I have found that Newton's ambivalent relation to mechanism and the mathematical abstraction of space and bodies, as expressed most visibly in his alchemical writings, recalls in surprisingly specific ways Milton's views on body and spirit. And in Newton's explicit statements of his motivations for opposing mechanical philosophies I have found circumstantial evidence for my excavation of Milton's motivations.

In saying that all created existents in Milton's world are material, I mean that they are bodily, and that there are no purely immaterial, incorporeal substances such as Plato or Descartes in their different ways affirmed; that angels, souls, and rational minds are not exceptions; and that there is an unbroken bodily chain linking stones and souls, anvils and angels, Earth and Heaven.

In placing both Milton and Newton in the context of vitalist theories in seventeenth-century England, I locate them as part of a new movement embracing

[10] See Timothy M. Harrison, "Every Living Thing": Life and Endeavor in Milton's "*Paradise Lost*," *Representations* 167 (2024): 33–63. While ultimately not adopting Harrison's definition, I have benefited immensely our continuing conversations on the definition of life. For Harrison's remarkable and philosophically learned study of the birth of consciousness in *Paradise Lost*, see his *Coming To: Consciousness and Natality in Early Modern England* (Chicago: University of Chicago Press, 2020).

[11] See my *Milton among the Philosophers*.

advances in minerology and mining, medical alchemy, and the philosophy of substance. Ku-ming Chang has distinguished "cosmic vitalism," which theorizes a universal spirit suffused through and animating the cosmos, from "immanent vitalism," which according to Chang "presumed a principle of life that was intrinsic to matter."[12] A familiar example of cosmic vitalism is that articulated by Anchises to Aeneas in the *Aeneid* 6.724–27: "First, know that heaven and earth and the watery plains, the moon's bright sphere and Titan's star, a spirit within sustains; in all the limbs mind moves the mass and mingles with the mighty frame."[13] Rather than a unitary universal spirit moving all things, the immanent vitalists described by Chang, and among whom I include Milton and Newton, argued for the possibility of body or extended stuff to be itself instinct with life and capable of self-motion, rather than being moved, as in the words Virgil gives to Anchises, by an extra-material spirit.

An understanding of spirit as rarefied body and body as congealed spirit is one of the most fascinating points of contact between Milton and Newton. At a time of vigorous debate among Neoplatonist, Cartesian, atomist, and Hobbesian views on the nature of corporeal substance and the nature and even existence of incorporeal substance, Milton argued for the unity of substance, for the existence of spirit and body along an unbroken continuum. In the *De Doctrina Christiana*, Milton explicitly repudiates a dualism of corporeal and incorporeal substances:

> man is an animate being [*animal*], inherently and properly one and individual, not twofold or separable—or, as is commonly declared, combined or composed from two mutually and generically different and distinct natures, namely soul and body—but... the whole man is the soul, and the soul is man: namely a body or substance which is individual, animated, sensitive, and rational.... [T]he idea of man's spirit being separated from his body, so as to exist somewhere apart, entire and intelligent, is not only not found anywhere in holy scripture, but also plainly clashes with nature and reason.[14]

The soul is not separable from the body; it is rather the "animated, sensitive and rational" body. Milton on the same manuscript page notes several scriptural instances in which "whatever is assigned to the body, the same is assigned to the

[12] Ku-ming (Kevin) Chang, "Alchemy as Studies of Life and Matter: Reconsidering the Place of Vitalism is Early Modern Chymistry," *Isis* 102, no. 2 (2011): 324.

[13] Virgil, *Eclogues, Georgics, Aeneid I–VI*, with an English translation by H. Rushton Fairclough, revised by G. P. Goold (Cambridge, MA: Loeb Classical Library, 1999); https://www.loebclassics.com/view/virgil-aeneid/1916/pb_LCL063.583.xml, accessed August 26, 2022. In the original:

> *Principio caelum ac terras camposque liquentis*
> *lucentemque globum lunae Titaniaque astra*
> *spiritus intus alit, totamque infusa per artus*
> *mens agitat molem et magno se corpore miscet.*

[14] Milton, *De Doctrina Christiana*, I.7; *OCW* 8.1:302–5.

soul," arguing that the terms "spirit" and "body" can be used to describe different components of the one substance: "whenever we speak of the body as though of a torso, then 'soul' means either the same as 'spirit,' or its less primary faculties: the vital or sensitive faculty, say." Milton acknowledges colloquial uses of *body*, *spirit*, and *soul*, but insists that there is no ontological divide between the aspects of ensouled existence that they signify.

In *De Doctrina Christiana*, Milton ties his argument against the separable existence of an incorporeal soul to scriptural evidence.[15] His concern, like Newton's in his arguments on substance, is largely theological. Their materialisms, as we will see, reflect their understandings of God and of the relation of the divine and human. At the same time, their thought is informed not only by theological but also natural philosophical trends in the seventeenth century.

In yoking "spirit" with "vital and sensitive facult[ies]," Milton points to the tenuously corporeal spirits of early modern physiology. Milton's Raphael in Book 5 of *Paradise Lost* describes the role of these tenuously corporeal spirits. According to Raphael, the "one first matter" is

> Endued with various forms, various degrees
> Of substance, and in things that live, of life;
> But more refined, more spiritous, and pure,
> As nearer to him placed or nearer tending
> Each in their several active spheres assigned,
> Till body up to spirit work, in bounds
> Proportioned to each kind. (*PL* 5.473–79)

Where Descartes had viewed the spirits as tenuous corporeal matter that communicated mysteriously with the incorporeal soul in the pineal gland, Milton's angel fills in the putative gap between mind and body in describing the sublimation of Galenic spirits into "intellectual spirits." Substance becomes "more refin'd, more spirituous, and pure" as it moves nearer to God (and by implication less refined, spirituous, and pure as it moves away from God).

Milton illustrates this dynamic scale of being with a plant simile that concludes with a startling identification of reason and the soul with highly refined corporeal spirits:

> So from the root
> Springs lighter the green stalk, from thence the leaves
> More airy, last the bright consummate flow'r
> Spirits odorous breathes: flow'rs and their fruit
> Man's nourishment, by gradual scale sublimed

15 He cites Lev. 5.2, 7.18, 7.20; Prov. 13.25, 25.25, 27.7; Isa. 29.8; 1 Sam. 24.11; and Ps. 7.5.

To vital spirits aspire, to animal,
To intellectual, give both life and sense,
Fancy and understanding, whence the soul
Reason receives, and reason is her being, (*PL* 5.479–87)

Galen's ascending triad of spirits—natural, vital, and animal—are translated into Raphael's, and Milton's, vital, animal, and intellectual.[16] The boundary, if there is one, between intellectual spirits and the soul, which receives reason (and "reason is her being") from those spirits, is permeable. In place of a gap that is inexplicably bridged in Descartes' pineal gland, Milton offers a material, corporeal continuity of nutrients, body, and soul.[17] The speech ends with an extraordinary simile in which plant metabolism is both metaphor for and dynamic instance of the conversion of body to spirit (i.e., of grosser body not into incorporeality but into increasingly tenuous and active body) by sublimation and rarefaction. Thinking is continuous with corporeality for Milton, rather than being the province of a separate, incorporeal substance. There is moreover no separable, incorporeal soul endowed with natural immortality. As we will see, related to this belief in the corporeality of the soul is his conception of an animate universe.

Phillip Donnelly has usefully pointed out that in earlier work I did not sufficiently and explicitly distinguish between what I meant by "matter" and "material" on the one hand and Aristotelian *prima materia* on the other.[18] Donnelly is correct that there is a distinction between the *matter* of Milton's chaos—that is, matter without the addition of forms—on the one hand and on the other the bodies or formed matter of *things* created from that disordered prime matter. Nevertheless, the matter of chaos in *Paradise Lost* and the original matter of creation in Milton's *De Doctrina Christiana* are not the qualityless, insensible *prima materia* of Aristotle. Milton's first matter is not insensible but rather disorderd or unordered. Milton describes in *Paradise Lost* a chaos that is sensible, a place of warring atoms, a place where one can find a "tumultuous cloud / Instinct with fire and niter," a place summed up as a "crude consistence" (2.936–37, 941). One might write off this evidence as the necessary shift of a poet obliged to describe what is after all insensible, were it not that he also describes the prime matter in the

[16] For a lucid account of Milton's relation to the Galenic tradition of corporal spirits, see Sarah Keleher, "'So all was cleared': Corrupted Fancy and Purgative Tears in *Paradise Lost*," in *Milton in Strasbourg*, eds Christophe Tournu, John Hale, and Neil Forsyth (Brussels: Peter Lang, 2022): 71–86.

[17] Recently, Jean-Luc Marion has argued cogently that Descartes proffered a term between *res extensa*, or bodies, and *res cogitans*, or mind, in the concept *meum corpum*, or "my body," which as the site of sensation or "passive thought" unites *res extensa* and *res cogitans* in the human being more intimately for Descartes than we have been used to thinking. Even if one grants Marion's argument, strict dualism nevertheless characterizes Descartes' world outside this intimate connection in the human being of *res extensa* and *res cogitans*. See Jean Luc Marion's *On Descartes' Passive Thought: The Myth of Cartesian Dualism* (Chicago: University of Chicago Press, 2018). I am indebted to Timothy Harrison for bringing this book to my attention.

[18] Phillip J. Donnelly, "'Matter' versus Body: The Character of Milton's Monism," *Milton Quarterly* 33, no. 3 (1999): 79–85.

discursive prose of *De Doctrina Christiana* not as insensible but as disordered; this matter is sensible but not yet beautiful: "It was a substance, and derivable from no other source than from the fountain-head of all substance; at first unarranged and disorganized, but afterwards God arranged it and made it beautiful" (*DDC* I.7; *OCW* 8.1:293).[19] While the terms *matter* and *material* can refer to Aristotle's or the scholastics' *prima materia*, Milton's prime matter is, unlike Aristotle's, not bereft of all qualities, including extension. It is, in other words, the corpuscularian prime matter that replaced Aristotle's qualityless *prima materia* as described by Pasnau in *Metaphysical Themes: 1274–1671*. In sum, *matter* in this book refers to extended corporeal stuff; *materialism* refers to the belief that all that is, is material extended stuff, and that souls and spirits are not incorporeal; and *animist materialism* refers to the position that life and self-activity are essential qualities of this material or corporeal substance, not of superadded incorporeal substance. For the sake of economy I will continue to use the terms *matter* and *material* with the stipulation that the terms serve to distinguish extended, corporeal substance from any proposed immaterial substance.

Going farther than Donnelly, Noel Sugimura has also challenged claims for Milton's animist materialism by arguing that Milton's "first matter" is in fact not extended corporeal stuff but insensible, unqualified *prima materia*.[20] The crucial passage in *Paradise Lost* is Raphael's speech on the chain of existents, which begins

> O Adam, one Almighty is, from whom
> All things proceed, and up to him return,
> If not depraved from good, created all
> Such to perfection, one first matter all,
> Endued with various forms, various degrees
> Of substance, and in things that live, of life. (5.469–74)

For Sugimura, created things are formed *out of* Milton's "one first matter," which she reads as Aristotelian *prima materia*. But Milton does not write "created all / ... [*from* or *out* of] one first matter." Raphael speaks of *things*, something one doesn't find in Aristotle's *prima materia*; Milton's chaos is a different realm of extended substance prior to the creation of ordered things. Disorganized or chaotic first matter is shaped into created *things*. All created things are composed of matter, Raphael suggests here, not in the metaphysical sense of *prima materia*, but in the natural philosophical sense of extended, sensible matter.

[19] "*Substantia erat, nec aliunde quàm ex fonte omnis substantiae derivanda, indigesta modo et incomposita, quam Deus postea digessit et ornavit*" (*OCW* 8.1:292).

[20] Noel Sugimura, *Matter of Glorious Trial: Spiritual and Material Substance in "Paradise Lost"* (New Haven, CT: Yale University Press, 2009), and "Milton and Matter," Oxford Handbook Topics in Literature (online edn, Oxford Academic, December 16, 2013), https://doi.org/10.1093/oxfordhb/9780199935338.013.11, accessed August 18, 2025.

Donnelly suggests the possibility and Sugimura asserts as fact a break in Milton between corporeal and incorporeal existents that share a first matter, inserting an ontological gap between "body" and "spirit" when Milton's Raphael speaks of a moment when "body up to spirit work[s]" (5.478). Donnelly writes that "although humans and angels are both rational substances differentiated out of matter, the former are necessarily corporeal (sensible) while the latter are potentially but not necessarily sensible in every respect."[21] What is it to be "potentially sensible"? It is not, of course, that Milton's angels are material only in the metaphysical sense of *prima materia*, for they are formed or organized. It is safe to assume that Milton regarded angels as sensible not only to each other but also to Adam and Eve. Milton goes out of his way to qualify the manner in which angels and souls might be thought to be "insensible" or "incorporeal." The key distinction is whether angels and souls are practically "insensible," because of the tenuousness of their matter and the relative bluntness of human senses, or absolutely insensible, because purely "incorporeal." The fact that Milton's angels can consume and digest corporeal food speaks loudly against pure incorporeality.

If Donnelly allows that angels are "potentially but not necessarily insensible," Sugimura is less cautious, arguing that Milton opens a real ontological gap between the corporeal body and incorporeal spirit or soul, as well as between human and angel. Like Donnelly, Sugimura stresses the equivocal use of "matter," which can refer both to the metaphysical prime matter, the paradoxical and insensible substance without the addition of forms, and to the matter of the things of physics, characterized by mass and dimension. Sugimura argues that Raphael's speech on the "one first matter all" starts off from the metaphysical prime matter before describing how this matter, endued with various forms, underlies a hierarchical chain of enformed bodies, before finally issuing in a "super-rarefied matter not found in nature."[22] She argues that while Raphael's metaphor "draws on how matter exists within the natural world (the subject of physics)," the tenor of the metaphor is not physical but metaphysical.[23] To the *prima materia*, the argument goes, are added forms that produce ontologically separate kinds: "all matter is stratified in a way that admits differences in *kind* (V.479), where 'kind' refers to the genus (viz., plants, animals, men, angels, etc.)."[24] Sugimura apparently stresses "kind" here to counter arguments that differences in Milton's creation are matters of degree rather than kind. But while "plants, animals, and men," or at least the bodies of men, are clearly material, Sugimura wants to suggest that angels and minds lie on the other side of an ontological gap. The leap is adumbrated, she argues, at the end of Raphael's plant simile quoted a moment ago:

[21] Donnelly, "'Matter' versus Body," 82r.

[22] Sugimura, "Milton and Matter," 7.

[23] According to Sugimura, "The angel's metaphor conveys a description of matter in distinctly metaphysical terms," "Milton and Matter," 6.

[24] Sugimura, "Milton and Matter," 4.

from the root
Springs lighter the green stalk, from thence the leaves
More airy, last the bright consummate flow'r
Spirits odorous breathes: flow'rs and their fruit
Man's nourishment, by gradual scale sublimed
To vital spirits aspire, to animal,
To intellectual, give both life and sense,
Fancy and understanding, whence the soul
Reason receives, and reason is her being. (5.479–87)

It is hard to see what more Milton could have done to deny a gap between corporeal and incorporeal. Sugimura's reading is ingenious but unwarranted. In moving from flower to the "spirits odorous" that it "breathes," she writes,

> The perfection of matter is no longer imagined as a tangible flower but as a wisp of a fragrance.... [T]he materializing force of the final image...—the flower—suddenly disappears.... [M]atter may remain, but it is not the same matter with which the image began. The matter at the end of the process—when man's "bodies may at last *turn all to Spirit*" (V.497; emphasis mine)—is distinct from the corporeal matter treated by physics (i.e., natural philosophy) at the beginning of his speech.[25]

Sugimura writes elsewhere of the "spirits odorous," in the same vein: "fruit smacks of the material, not spiritual world, and is in sharp contrast to the thing imaged in...the ascent—...the 'bright consummate floure,' whose fragrance wafts toward immateriality."[26] This last clause, "whose fragrance wafts toward immateriality," has a lot of heavy lifting to do. Fragrances, while carried by very tenuous effluvia, are neither immaterial nor insensible. While they are in common speech less material than roots or stems, in metaphysical terms they are not divided from roots or stems by a leap across an ontological divide. Eve will herself be "Veiled in a cloud of fragrance" when Satan espies her in Book 9 (9.425). Odors may *waft toward immateriality* in the relative sense of being less solid, but not in any absolute sense. There are similar conflations of tenuous materiality and putative immateriality as the analysis proceeds. The "stuff of angels" becomes, as quoted above, "a super-rarefied matter not found in nature," and the ascent from human to angel involves a "transmogrification," an "ontological leap of substance," occurring now not in the progression from root to stalk to leaf to flower, but in a moment when, as Raphael suggests (5.483–501), human beings might "participate" with angels (or, Sugimura suggests, trade materiality for immateriality).[27] But if there

[25] Sugimura, "Milton and Matter," 6.
[26] Sugimura, *Matter of Glorious Trial*, 47.
[27] Sugimura, "Milton and Matter," 7, 10.

is an ontological leap into immateriality in a natural philosophical sense, it is difficult to see how Milton's angels can have "proper shapes" (*PL* 3.634 and 5.276).

The argument for an ontological leap would be persuasive if every change in the composition of an embodied substance, if, for example, every attenuation of food in digestion, every budding of flower from stalk, comprised an "ontological metamorphosis," but the argument instead seems to be that this unique metamorphosis occurs as a gap is crossed between matter as understood by physics and some incorporeal and absolutely rather that practically insensible spirit or incorporeal substance. One can find in *Paradise Lost*, as both Donnelly and Sugimura point out, passages differentiating *corporeal* and *incorporeal*, but these passages indicate relative materiality in physical terms. Relative to wood, fire is immaterial, but it is still composed of minute bodies.

I suggest that one will look in vain in Raphael's speech for a moment of ontological discontinuity. There is no ontological leap, though there is an increasing attenuation and purification of matter, as corporeal food is digested and sublimed (a physical process) into increasingly tenuous corporeal spirits, which themselves "give," Milton's term, faculties up to and including "understanding," from which "the soul / Reason receives, and reason is her being" (note that the verb *to be* is in the stressed position). The soul does not have a logically prior being that receives faculties from sublimed physical matter. What it receives *is* her being. Milton's formulation is designed to thwart a belief that intellectual spirits, his invented term, are discontinuous with the vital and animal spirits that no one would claim to be incorporeal. Raphael's promise to Adam and Eve that their "bodies may at last turn all to spirit" (5.497) comes at the end of a speech in which the keynote is continuity in the chain of being, not ontological discontinuity. "[B]ody" could not "up to spirit work" if there were an ontological chasm to be crossed (5.478).

There seems no warrant in the text for a disjunctive ontological metamorphosis; instead, the claim seems based on the assumption rather than demonstration that the "matter" of angels must be material only in the metaphysical sense and not in the natural philosophical sense. The poem, on the other hand, indicates, and Sugimura acknowledges, that Milton's angels can digest and assimilate corporeal food, a marked departure from the scholastic tradition that Sugimura sees as operative in Milton's thinking on this question, a tradition that viewed the angels who eat in Scriptural accounts as doing so "in mist, the common gloss / Of theologians" (5.435–36), that is, as gathering temporary bodies of air that only seem to eat. Milton's angels digest with "concoctive heat" (5.437).

Vitalism and Active Laws in Newton

Like Milton, Newton rejects the need for a separate incorporeal substance to account for life and thought, and he views the universe of matter as alive. Seeing

Newton as a mechanist who proposed a clockwork universe has become increasingly untenable as Newton's extensive theological and alchemical writings have come to light. The great natural philosopher saw God's direct hand in the creation and continued operation of the cosmos, and he disputed the Cartesian divide between thinking and extended substances explicitly even as Milton had implicitly, in a passage in his manuscript "De gravitatione et æquipondio fluidorum" ("On the gravity and equilibrium of fluids") to which I will return in a moment. And while it may be surprising to those who know him only or mainly as the architect of the laws of force and inertia, Newton thought of the cosmos as animate; "all matter duly formed," he writes in the Draft Queries to his *Opticks*, "is attended with signes of life."[28] Newton, as I will demonstrate below, resembles Milton not only in his general conception of the life of things and the corporeality and mortality of the soul, but also in his imagery and his language of universal life.

For Newton, theories of mechanically driven passive matter offer only a partial window onto nature. Taken as the whole story, mechanism falsifies our understanding of nature. Where Cudworth allowed a severely limited place for mechanism in a world largely moved by incorporeal substance, Newton speculated freely on the organic and formative power of living and self-moving corporeal substance. Newton refers, as we will see, to the laws of inertia and force as "passive laws," and he insists on the presence of active laws.[29] He referred to these active *laws* sometimes as active *principles* or, in at least one case, active *spirits*,[30] and when he does so, particularly in his alchemical writing but also in his "Queries" to the *Opticks*, the result is a picture of the world akin to Milton's version of animist materialism. Newton, like Milton, rejected any form of simple dualism or merely mechanist materialism. For Newton, body emanates from spirit, not as a separate substance but as contained in spirit. Having argued that God contains extension—a defining quality of bodies—within himself, on the principle that a God that did not contain extension within himself could not have created extension and extended bodies, Newton describes the unbroken connection of minds and bodies:

> if extension is eminently contained in God, or the highest thinking being, certainly the idea of extension will be eminently contained within the idea of thinking, and hence the distinction between these ideas will not be so great but

[28] "Draft version of Queries to the *Optics*," Cambridge University Library Additional MS 3970.3, fol. 619^r, quoted from http://www.newtonproject.ox.ac.uk/view/texts/normalized/NATP00055, accessed August 18, 2025.

[29] "Draft version of The Queries," 619^r; J. E. McGuire and P. M. Rattansi point to the significance of this passage in "Newton and the Pipes of Pan," *Notes and Records of the Royal Society of London* 21, no. 2 (1966): 118.

[30] In the "Draft version of Queries to the *Optics*," Newton suggests that "all bodies...abound with a very subtile active potent elastic spirit," fol. 235^r, accessed August 7, 2025.

> that both may fit the same created substance, that is, but that a body may think, and a thinking being extend [*hoc est corpora cogitare vel res cogitantes extendi*].[31]

Newton's assertion of the continuity of thinking and extension, the defining characteristics of René Descartes' two separate substances, echoes Milton's description in the *De Doctrina Christiana* of the relation between spirit and body in God's creation of the universe:

> It is objected that body cannot emanate from spirit. I reply, much less can it emanate from nothing; for Spirit, being the more excellent substance, virtually (as they say) and eminently contains within itself what is undoubtedly the inferior substance—just as the spiritual and rational faculty contains the corporeal one, that is, the sentient and vegetative faculty.[32]

Milton specifies in the *De Doctrina Christiana* that "God in his simplest nature is spirit/*Deus sit natura simplicissima* ***Spiritus***" (*DDC* I.2; *OCW* 8.1:34, 35) and that no other than God can be the "material cause" (*DDC* I.7; *OCW* 8.1:293).[33] Milton's God contains material cause as spirit eminently contains body. Given that the defining quality of body (or matter) is extension, when Newton writes that "extension is eminently contained in God," he apparently agrees with Milton on body emanating from spirit, and by extension from God.

In concluding "that a body may think, and a thinking being extend," Newton like Milton not only tied the origin of the world to the Creator but also tied life, consciousness, and activity to body without requiring the addition of incorporeal spirit.[34] Both also imply the inter-convertibility of spirit or mind and body. If they are manifestations of one created substance, then alterations of that substance account for both apparently unthinking things (such as elm trees) or even apparently inanimate things (such as stones) on the one hand and soul or mind on the other. Newton's picture here also parallels that of Anne Conway, a protégé of Henry More who broke with her mentor's dualism in articulating her own

[31] "On the Gravity and Equilibrium of Fluids," in A. Rupert Hall and Marie Boas Hall, eds., *Unpublished Scientific Papers of Isaac Newton* (Cambridge: Cambridge University Press, 1962), 143 (Latin original, 109). While Betty Jo Teeter Dobbs dates the manuscript to 1684–85 (*The Janus Faces of Genius: The Role of Alchemy in Newton's Thought* [Cambridge: Cambridge University Press, 1991], 139–41), A Rupert Hall and Marie Boas Hall suggest a date in the 1660s.

[32] Milton, *De Doctrina Christiana*, I.7, *OCW* 8.1:295.

[33] "*Materialis...causa erit aut Deus aut nihil*" (*OCW* 8.1:292).

[34] While Newton's opposition to dualist claims for a separate incorporeal substance might at first glance seem to draw him toward Spinoza's monism, he vigorously opposes Spinoza, on theological grounds, for his identification of God with nature, on the grounds that it delivered God and creation over to blind necessity; he attacks Spinoza implicitly in the General Scholium to the *Principia* for denying final causes and the design argument. See in this connection Eric Schliesser, "Newton and Spinoza: On Motion and Matter (and God, of Course)," *The Southern Journal of Philosophy* 50, no. 3 (2012): 436–58.

version of alchemically inflected vitalist monism in *The Principles of the Most Ancient and Modern Philosophy*. Conway writes that "the distinction between spirit and body is only modal and incremental, not essential or substantial [*modalis & gradualis, non essentialis vel substantialis*]."[35] It also recalls the angel Raphael's speech, discussed above, on the one first matter in *Paradise Lost* (5.469–90), in which "body" can "up to spirit work" (78).

Newton's rejection of the view that body and spirit are separate substances fits his refusal to treat mechanical causation as sufficient to account for all corporeal phenomena. As one would expect from the architect of the laws of motion, Newton did recognize that mechanical causation could account for some effects, but he insists that a full account of nature must embrace another set of causes. He writes in the 1675 draft version of the "Queries" to the *Optics* that

> Matter is a passive principle & cannot move it self. It continues in its state of moving or resting unless disturbed. It receives motion proportional to the force impressing it. And resists as much as it is resisted. These are passive laws & to affirm that there are no other is to speak against experience. For we find in our selves a power of moving our bodies by our thought[.] Life & will are active Principles by which we move our bodies, & thence arise other laws of motion unknown to us.[36]

Dmitri Levitin points out that Newton's argument here is directed against Cartesian mechanism, with its assumption that motions in the physical world of *res extensa* are caused by the prior physical motions of objects (and Descartes' vortices[37]). In insisting on non-mechanical active laws, Newton is calling out what he sees as the atheistic implications of Cartesian mechanism.[38] The view that mechanical laws might explain the origins and maintenance of the cosmos,

[35] Anne Conway, *The Principles of the Most Ancient and Modern Philosophy*, ed. and trans. Allison P. Coudert and Taylor Corse, *Cambridge Texts in the History of Philosophy* (Cambridge: Cambridge University Press, 1996), 40 (Chapter 6). I have modified the recent Coudert and Corse translation, changing "and" to "or" in the final phrase. The book, which Conway composed in English shortly before her death in 1679, was first published in a Latin translation in 1690, and then in a translation back into English in 1692. For the Latin and the first translation, see Anne Conway, *The Principles of the Most Ancient and Modern Philosophy*, ed. Peter Loptson (The Hague: Martinus Nijhoff, 1982), 102. I compare Conway's monist vitalism with Milton's in *Milton among the Philosophers*, 117–36.

[36] Query 23 in the "Draft version of Queries to the *Optics*," fol 619^r, accessed August 7, 2025.

[37] Cartesian cosmology posits a plenum (i.e., voidless space) of circling bands or *vortices* of three material elements: (1) "violently agitated" matter that is "divided into particles of indefinite smallness" and that makes up the sun and stars, (2) minute and to our eyes invisible particles that make up the heavenly ether, and (3) "bulkier" particles "less suited for motion" that make up planets and comets," *Principles of Philosophy*, III:52, in *The Philosophical Writings of Descartes*, trans. John Cottingham, Robert Stoothoff, and Dugald Murdoch, 2 vols (Cambridge: Cambridge University Press, 1984–85), I:218.

[38] Dmitri Levitin, *The Kingdom of Darkness: Bayle, Newton, and the Emancipation of the European Mind from Philosophy* (Cambridge: Cambridge University Press, 2022), 657–58.

such as Descartes proposed (even as a hypothetical counterfactual to divine creation), threatened to make God unnecessary.[39]

Newton in the same manuscript argues that passive or mechanical laws are insufficient to account for physical phenomena:

> The Vis inertiæ is a passive principle by which bodies persist in their motion or rest, receive motion in proportion to the force impressing it, & resist as much as they are resisted. By this principle alone there could never have been any motion in the world. Some other Principle was necessary for putting bodies into motion, & now they are in motion some other Principle is necessary for conserving the motion.[40]

This "Vis inertiæ" is the property of matter as mere body in the absence of the operation of active corporeal principles. While we can observe that colliding billiard balls follow mechanical laws in their motions, we can also observe both (1) how our wills can initiate motion in our bodies and (2) how the balls on a billiard table come to rest. The introduction of active principles or laws is necessary not only for the origin of the cosmos but for its conservation or motion and thus the continued operation of the cosmos. In the absence of the impetus of active principles, friction in the system would eventually lead to inert stasis. Newton was intent on discovering how non-mechanical causes, which he called "active laws" and "active principles" interchangeably, functioned:

> And since all matter duly formed is attended with signes of life & all things are framed with perfect art & wisdom & Nature does nothing in vain; if there be an universal life & all space be the sensorium of a thinking being who by immediate presence perceives all things in it as that which thinks in us perceives their pictures in the brain: the laws of motion arising from life or will may be of universal extent.[41]

As we move our bodies, so God can move objects in the world. God, moreover, can create corporeal active principles that initiate change and motion. Newton devoted enormous time and energy to discovering the manner in which non-mechanical and material active causes operated in the world. These active causes were at work most obviously in organisms, as one would expect from his analogy between human will and motion and God's will as the origin of motions. They

[39] See René Descartes, "Discourse on the Method," Part 5, in *The Philosophical Writings of Descartes*, ed. and trans. John Cottingham, Robert Stoothoff, and Dugald Murdoch, 2 vols (Cambridge: Cambridge University Press, 1984–85), 1:133–34. See similar arguments in the sixth chapter of Descartes' *Le Monde* and the *Principles of Philosophy* III, 45 (Cottingham 1:91,256).

[40] Draft Version of the Queries, 3, 254r, accessed January 18, 2024.

[41] Query 23 in the "Draft version of Queries to the *Optics*," fol 619^{r}, accessed August 7, 2025.

were also at work in chemical and alchemical reactions, instances included in or analogous to organic process, as we will see in the next chapter. While Newton famously posited that he merely charted the mathematical laws governing gravity and made no hypotheses concerning gravity's cause, he also suggested that God caused gravity either directly or indirectly through active principles. Newton vacillated on the question of the cause of gravity. At times he treated gravity as a mechanical phenomenon. In the alchemical manuscript "Of natures obvious laws and processes in vegetation," written in or shortly before 1672, Newton included gravity among phenomena explicable in terms of mechanical (or "passive") laws: "Natures actions are either vegetable or purely mechanicall (grav. Flux[es]. meteors. vulgar Chymistry)[.]"[42] By the mid-1670s, however, he was drawn to the idea that gravity was an effect of active principles in the interplanetary aether, as in this letter of December 7, 1675 to Milton's friend Henry Oldenburg: "so may the gravitating attraction of the Earth be caused by the continuall condensation of some other such like æthereall Spirit, not of the maine body of the flegmatic æther, but of something very thinly & subtily diffused through it, perhaps of an unctuous or Gummy, tenacious & Springy nature, and bearing much the same relation to æther, wch the vitall æreall Spirit... does to Air."[43] The active principles are material but, unlike what Newton thinks of as mere body, they initiate motion. Newton relates the action of gravity to the electric and magnetic phenomena that he traces to active principles in the form of "Electric & Magnetic effluvia."[44] He also relates the subtle active principles perhaps responsible for gravity to the spirits that govern organic processes. In the next chapter we will see how Newton endeavored through alchemical reading and experimentation to understand the existence and activity of these spirits.

Where the Cambridge Platonists Cudworth and More posited a clear divide between entirely passive corporeal substance and active incorporeal substance, Newton like other experimentalists was interested in exploring the existence of material active principles in bodies. According to Ernan McMullin, "there can be no doubt that Newton *does* speak at times of a continuous spectrum of beings, from the level of pure spirit to that of totally passive matter."[45] That these entities are located along a continuous spectrum, rather than standing on opposing sides of an ontological gap between the corporeal and incorporeal, sets Newton apart from More and Cudworth, and brings him closer to Milton, for whom all that

[42] "Of natures obvious laws," folio 5^r, http://webapp1.dlib.indiana.edu/newton/mss/norm/ALCH00081, accessed August 11, 2018; I follow Betty Jo Teeter Dobbs' dating of the manuscript, in *The Janus Faces of Genius: The Role of Alchemy in Newton's Thought* (Cambridge: Cambridge University Press, 1991), 15.

[43] *The Correspondence of Isaac Newton*, ed. H. W. Turnbull, X vols (Cambridge: Cambridge University Press, 1959–77), I:365; quoted also in Dobbs, 102.

[44] *Correspondence*, I:364.

[45] Ernan McMullin, *Newton on Matter and Activity* (Notre Dame, IN: University of Notre Dame Press, 1978), 19.

exists is arrayed along a spectrum of "one first matter," from the gross to the "more refined, more spirituous and pure."

Newton's friend More did blur the boundary between the corporeal and the incorporeal in proposing, in *The Immortality of the Soul* and "The True Notion of a Spirit," that incorporeal substances share a quality commonly viewed as defining corporeal substances—extension[46]—and the editors of the *Acta Eruditorum* did suggest that Newton's spirits were "perhap[s the same with the Hylarctick principle of D. H. More."[47] But Newton at this point denies that he has specified whether his "spirits" are material or immaterial. Elsewhere in his work, however, and particularly in his alchemical writings, he speculated on the existence of material active principles.

Newton spoke of these active principles sometimes as laws[48] and sometimes as tenuously corporeal spirits. In a particularly interesting manuscript mentioned briefly above, "Of natures obvious laws," he describes this spirit as the "material soul of all matter":

> This is the subtil spirit which searches the most hiden recesses of all grosser matter which enters their smallest pores & divides them more subtly then any other materiall power what ever. (not after the way of common menstruums by rending them violently assunder etc) this is Natures universall agent, her secret fire, the onely ferment & principle of all vegetation. The material soule of all matter which being constantly inspired from above pervades & concretes with it into one form & then if incited by a gentle heat actuates & enlivens it.[49]

This spirit for Newton is the spring of organic organization, nutrition, growth, and decay. Unlike More's "spirit of nature" and Cudworth's "plastic nature," which it resembles in function, Newton's "subtile spirit" is not incorporeal but rather the "material soul of all matter." It is "inspired" and "incited" not by an incorporeal

[46] See Henry More, *The Immortality of the Soul* (1659), rpt., in *A Collection of Several Philosophical Writings*, 2 vols (1662; facs. rpt. New York: Garland, 1978), 3, Preface; and "The Easie, True, and Genuine Notion and Consistent Explication of the Nature of a Spirit," rpt., in Florence Isabel MacKinnon, *Philosophical Writings of Henry More* (New York: Oxford University Press, 1925), sec. xxi, 212.

[47] See the passage in Newton's notes on his dispute with Leibniz: "In the Acta Eruditorum for March 1714 p. 140 and 141, it's said that Mr Newton denies that the cause of gravity is the [sic] mechanical, & that he laies down a certain new Hypothesis concerning a subtile spirit pervading the pores of bodies (perhaps the same with the Hylarctick principle of Dr H. More) & this spirit is represented of less value then Hypotheses are unless it be the Æther or subtile matter of the Cartesians. And yet Mr Newton has no where denyed that the cause of gravity is Mechanical nor affirmed whether that subtile spirit be material or immaterial," Cambridge University Library MS Add. 3968, ff. 594r–619v, Cambridge University Library, 125^r^, https://www.newtonproject.ox.ac.uk/view/texts/normalized/NATP00385, accessed October 19, 2022.

[48] As he does in Query 31 to the 1706 Latin edition of his *Optics*.

[49] "Of natures obvious laws," folio 3^v^, accessed August 5, 2025.

spirit but by heat. Newton speculated that the active spirits might be akin to or identical to light:

> This spirit perhaps is the body of light 1 becaus both have a prodigious active principle both are perpetuall workers 2 because all things may be made to emit light by heat, 3 the same cause (heat) banishes also the vitall principle. 4 Tis suitable with infinite wisdom not to multiply causes without necessity 5 Noe heat is so pleasant & *illeg.* [perhaps "benign"] as the suns. 6 light & heat have a mutuall dependance on each other & noe generation without heat. heat is a necessary condition to light & vegetation.* 6 Noe substance so indifferently, subtily & swiftly pervades all things as light & noe spirit searches bodys so subtily percingly & quickly as the vegetable spirit. [*heate exites light & light exites heat, heat excites the vegetable principle & that increaseth heat.][50]

The subtlety of light particles allowed them to penetrate even dense matter, and its heat made light the ideal alchemical agent of quickening, vegetable growth, and putrefaction.[51] Light also played a crucial role for Conway, who complains that mechanists stop at the extension and impenetrability of bodies: "They do not go beyond the husk and shell, nor ever reach the kernel," which is "spirit or life and light." This light and life for Conway are the capacity for knowledge and feeling, which she extends even to "dust and sand…through various successive transmutations."[52]

Significantly, given Newton's insistence in the *Principia* on excluding hypotheses on the causes of the forces that he describes mathematically, in a brief final paragraph to the General Scholium of the second and third editions, Newton does offer a hypothesis on the material foundation of those forces:

> A few things could now be added concerning a certain very subtle active spirit pervading gross bodies and lying hidden in them (*spiritu quodam subtilissimo corpora crassa pervadente, & in iisdem latente*); by its force and actions, the particles of bodies attract one another at very small distances and cohere when they become contiguous; and electrical [i.e., electrified] bodies act as greater distances, repelling as well as attracting neighboring corpuscles; and light is emitted, reflected, refracted, inflected, and heats bodies; and all sensation is excited, and the limbs of animals move at command of the will, namely, by the

[50] "Of natures obvious laws," folio 4^r, accessed August 7, 2025; note that "exites" in the note appended to item 5 in this passage is Newton's spelling of "excites."

[51] Iliffe discusses the role of light as "a source of agency for Newton" in the *Opticks* as well as in alchemical texts, in "Abstract Considerations: Disciplines and the Incoherence of Newton's Natural Philosophy," *Studies in History and Philosophy of Science* 35 (2004): 427–54 (quoted passage from 449).

[52] Conway, *Principles*, 66 (Chapter 9).

> vibrations of this spirit being propagated through the solid fibers of the nerves from the external organs of the senses to the brain and from the brain into the muscles. But these things cannot be explained in a few words; furthermore, there is not a sufficient number of experiments to determine and demonstrate accurately the laws governing the actions of this spirit.[53]

Newton's most celebrated and revolutionary text ends with hinting at possibilities about which the text is otherwise resolutely agnostic. That agnosticism returns in the last sentence, as Newton demurs from supplying the greater elaboration impossible "in a few words" and acknowledges a lack of sufficient experimentation to test the hypotheses that he earlier in the General Scholium refuses to frame or feign. The hints, however, remain, and they echo possibilities and probabilities that he pursues in his unpublished manuscripts. Subtle, and given what he says elsewhere, material active spirits pervade and move grosser and more inert bodies. This seems to be the "material soule of all matter" found in the "most hiden recesses of all grosser matter" and the "prodigious active principle" of the manuscript "Of natures obvious laws." The "few things" gestured toward but not elaborated at the end of the "General Scholium" open a passage between, on the one hand, the masterwork determinedly agnostic in ignoring physical causes for the phenomena and forces that it explains mathematically and, on the other hand, the speculation about those physical causes in unpublished manuscripts.

That Milton consistently and Newton at times held vitalist ideas about matter is clear from their writings. The question that remains is *why*. What led them to ascribe life to what might seem to most then as inert, passive body? Why did Newton, as we will see later in this chapter, describe the cosmos in terms uncannily similar to Milton's picture in *Paradise Lost* of the solar system as a living, metabolizing body? These are questions that I will turn to in the next section.

Vitalism and the Reactions to Mechanism and Incorporealism

If one thinks of the seventeenth century as the time when older, naïve animist conceptions of nature were conquered by new rational and mechanist understandings of nature, Milton's and Newton's moves will seem anomalous. Similarly, if one thinks of as exhaustive the alternative metaphysical models of substance dualism (whether Platonist or Cartesian) or a monism of passive material bodies

[53] Newton, *Principia*, 943–44. The Latin original is quoted from the Newton Project's etext of the third book of the 1726 third edition *Principia* (identical to the passage in the second edition of 1713), https://www.newtonproject.ox.ac.uk/view/texts/diplomatic/NATP00303, 530, accessed December 8, 2023.

(as in Hobbes), Milton and Newton will again appear to be oddities. But these ways of looking at seventeenth-century English natural philosophy and metaphysics are misleading. Vitalist conceptions of substance were if anything dominant among English natural philosophers, not least because of a strong reaction to models of thoroughgoing mechanist causation exemplified by the dualist Descartes' understanding of *res extensa* or the monist Hobbes' mechanical and arithmetical understanding of all phenomena, whether mental or physical.

John Rogers has influentially argued for the brief flourishing of a "vitalist moment" in England in the middle of the seventeenth century.[54] This "vitalist moment," however, was not brief. Vitalist and animist ideas characterized the work of writers and natural philosophers from the beginning of the century to the end and beyond. In some cases, one can trace the influence of Paracelsus' sixteenth-century theory of animate spirits. This was the case early in the seventeenth century with Francis Bacon, who posited a continuity of inanimate and animate corporeal spirits, with the inanimate being so because "cut off" and dispersed in grosser bodies.[55] These inanimate spirits, however, have the potential to be "kindled" into animate spirits in the right circumstances.[56] While Bacon labels these lower spirits as "inanimate," he suggests that they contain within themselves the potential for animation: "*if it* [the inanimate spirit] *has the resources* for joining itself up and through that for exercising and enjoying its own nature, then indeed it kindles and conducts itself *according to its capability*" (my emphasis).[57] The seventeenth century, as we will see, offers many examples of philosophers positing a spectrum of matter ranging from the inanimate through the potentially animate to the animate, and philosophers seeking the source of life in neither an incorporeal soul nor a mechanical arrangement of passive corpuscles.

The rationales for the claims for self-active matter were often moral and theological. Cavendish, for example, ties her vitalism to the possibility of freedom of the will and divine justice. She makes explicit a connection I have argued is implicit in Milton, between his animist materialism and his defense of free will (thus supporting his defense of divine justice) from the determinism consequent on mechanist philosophies of matter. Another powerful theological motivation behind claims for self-active matter, one that would be of particular interest to Newton, was the defense of divine omnipotence, a point explicitly made by Cavendish, by the Christian virtuosi such as Robert Hooke and Robert Boyle, and by John Locke. Milton and Newton shared misgivings about thoroughly

[54] John Rogers, *The Matter of Revolution: Science, Poetry, and Politics in the Age of Milton* (Ithaca: Cornell University Press, 1996).

[55] I outline Bacon's natural philosophy of spirits in *Milton among the Philosophers*, 111–14.

[56] Francis Bacon, *Novum Organum* II, xl, in *The Works of Francis Bacon*, eds James Spedding, Robert Leslie Ellis, and Douglas Denon Heath, 14 vols (London: Longman, 1858–64), 4:196. For the language of "kindling," see *De viis mortis*, fol. 26v, in Graham Rees, *Francis Bacon's Natural Philosophy: A New Source* (Chalfont-St. Giles: The British Society for the History of Science. 1984), 158–59.

[57] *De viis mortis*, fol. 26v.

mechanist understandings of causation with a broad array of English philosophers and experimentalists.

The great advantage of purely mechanist understandings of causation in the natural world is availability to common sense. One sees on the macroscopic level interacting corporeal motions. A ball on the billiard table or bowling green hits another at a particular velocity from a particular angle, and from experience one can predict the motion of the struck ball. In a mantel clock descending weights cause the rotation of gears, which in turn cause the dial's hand to advance. It is easy to imagine that similar interactions occur among atoms or parcels of a material plenum on a scale invisible to the senses, leading to sensible phenomena. The continued motion visible in objects from the terrestrial to the celestial is explicable in terms of inertia, the principle that an object would remain in uniform motion or rest unless or until an interaction with another object.

There were, however, objections to models of causation depending on the linked movements of passive, inertial bodies.[58] Those indebted to scholastic philosophy objected that simple contact could not explain how motion, which is a mode of a body, could be transferred to another body. Challenging the mechanism reigning in Cartesian *res extensa*, Henry More wrote to Descartes in 1649 that "I feel more disposed to believe that motion is not communicated, but that from the impulse of one body another body is so to speak roused into motion, like the mind to a thought on this or that occasion.... [M]otion bears the same relation to a body as a thought does to the mind: neither is received into the subject, in fact, but both arise from the subject in which they are found."[59] While hearing the thoughts of others may give rise to the hearer's thoughts, they remain hers, despite resembling what was heard in whole or in part. The mind is the substance that thinks. In the same way, for More, motion, either potential or actual, belongs to the moving material object itself.

Cavendish, who in 1663 corresponded with More and sent him copies of her works—and who later criticized his dualist account of substance—agreed with More on this point.[60] Motion is not transferred from body to body; instead the motion of one body is the occasion for the independent motion of the second. The idea fits better with Conway's vitalist materialism than with More's dualism;

[58] Levitin lays out the objections of seventeenth-century natural philosophers, such as William Harvey, Thomas Willis, and Robert Boyle, to what they saw as the mechanist reductionism of ancient atomism; they viewed mechanism as unequal to the challenge of explaining organic phenomena, and they pointed to chemical reactions rather than collision mechanics (*Ancient Wisdom*, 372–97).

[59] Henry More, letter to Descartes, July 23, 1649, as quoted in *Problems of Cartesianism*, eds Thomas M. Lennon, John M. Nicholas, and John W. Davis (Montreal: McGill-Queens University Press, 1982), 211. John Henry quotes the beginning of the passage in "Occult Qualities and the Experimental Philosophy: Active Principles in Pre-Newtonian Matter Theory," *History of Science* 24 (1986): 336. Henry's argument informs my discussion of seventeenth-century objections to mechanism.

[60] Lisa T. Sarasohn, *The Natural Philosophy of Margaret Cavendish: Reason and Fancy during the Scientific Revolution* (Johns Hopkins University Press, 2010), 135–41.

in the latter it is hard to understand how passive, inert matter could be "roused into motion." For Conway, all activity arises from the self-active matter of the object changing or moving, not from the object with which it comes in contact. Cavendish writes in the 1666 *Observations upon Experimental Philosophy*,

> there is no such difference as the learned make between patient and agent, when they call the exterior occasional cause, as (for example) fire, "the agent"; and the combustible body, "the patient"; for they conceive that a body thrown into fire, acts nothing at all, but only in a passive way suffers the fire to act upon it, . . . those bodies on which they suppose fire doth work, change according, not to the fire's, but their own inherent figurative motions: . . . and if self-motion be corporeal, then every part of nature must of necessity move by its own motion; *for no body can impart motion to another body, without imparting substance also*; . . . because matter and motion are but one thing.[61]

The fuel is not passively consumed by the fire; the fuel actively imitates in its own fashion the motions of the fire, which according the Cavendish the fuel *perceives*. In her 1664 *Philosophical Letters* she offers the example of how wax takes the form of a seal: "the seal doth not give the wax the print of its own figure, but it is the wax that takes the print or pattern from the seal, and patterns or copies it out in its own substance, just as the sensitive motions in the eye do pattern out the figure of an object."[62] For Cavendish, the same is true for colliding billiard balls. The stationary ball *perceives* the motions of the approaching ball and, patterning its own motions on those external motions, moves itself. In both More and Cavendish, there is no transfer of quality (motion or figure) from one substance (individual object) to another, such as would violate the scholastic principle that qualities belong to bodies and cannot be transferred to another without a transfer of substance.

If thoroughgoing mechanist theories of causation by collisions of inert matter were problematic from a scholastic philosophical perspective, they posed an even greater theological threat. Explaining natural phenomena by concatenations of passive matter came dangerously close to making God's creative agency unnecessary, an implication that as we have seen troubled Newton. Epicurean philosophy, as outlined in the *De rerum naturae* of Lucretius, posited the origin of the world from an initial swerve of eternal atoms hitherto falling in parallel lines.[63]

[61] Cavendish, *Observations*, II.20 (my emphasis), ed. Eileen O'Neill, Cambridge Texts in the History of Philosophy (Cambridge: Cambridge University Press, 2001), 234–35.

[62] Margaret Cavendish, *Philosophical Letters, or, Modest Reflections upon Some Opinions in Natural Philosophy Maintained by Several Famous and Learned Authors of this Age, Expressed by Way of Letters by the Thrice Noble, Illustrious, and Excellent Princess the Lady Marchioness of Newcastle*, Letter 32 (London, 1664), 105; Early English Books Online, ProQuest doc. 2240875599, accessed February 22, 2024.

[63] Lucretius, *De rerum natura*, 2.216–50.

Lucretius' poem is not atheistic, but its gods are distant and unconcerned with our affairs and our world, which they did not create. The seventeenth-century revivers of Epicurean atomist philosophy, Pierre Gassendi and Walter Charleton, were careful to distance themselves from the Epicurean/Lucretian argument against divine creation. Charleton, in a 1654 work the title of which signals reliance on Gassendi, the *Physiologia Epicuro-Gassendo-Charletoniana*, denies the Epicurean tenet of the eternity of atoms, and argues that God created atoms *ex nihilo*, and "invigorated or impregnated them with an Internal Energy or Faculty Motive, which may be conceived the First Cause of all Natural Actions, or Motions."[64] What Charleton calls atoms' "Internal Energy" or, elsewhere, the "faculty of self-motion"[65] serves the same purpose as Milton's active, living matter and Newton's "active principles," opening a space for divine creativity and ordering in what otherwise might seem the blind, mechanical origin and operation of the universe. The theorization of occult active principles (occult not because magical but because their operation inside matter was not directly observable) was common not only to metaphysicians but also to experimentalists such as Boyle and Hooke and Newton himself.

The specter of mechanistic atheism was embodied for Milton's and Newton's contemporaries by Thomas Hobbes, dubbed the "monster of Malmesbury."[66] In Hobbes' mechanist materialism, not only physical but also mental phenomena result from and are identical with motion of particles, and "*Conceptions*...are nothing *really*, but *motion* in some internal substance of the *head*."[67] The implication, which Hobbes embraces, is that "every act of mans will, and every desire, and inclination proceedeth from some cause, and that from another cause, which causes in a continuall chaine (whose first link [is] in the hand of God the first of all causes) proceed from *necessity*."[68] From this perspective, any assertion of the will's freedom from the iron chains of causal necessity is not merely unpersuasive, it is meaningless, an abuse of speech: "If a man should talk to me of a *round*

[64] Walter Charleton, *Physiologia Epicuro-Gassendo-Charletoniana, or a Fabrick of Science Natural, upon the Hypothesis of Atoms* II.iv (London, 1654), 126; Early English Books Online, ProQuest doc. 2240941769, accessed February 22, 2024.

[65] Walter Charleton, *The Darknes of Atheism Dispelled by the Light of Nature* (London, 1652), 47; Early English Books Online, ProQuest doc. 2264194095, accessed February 22, 2024.

[66] For the reaction to Hobbes' supposed atheism, see Samuel Mintz, *The Hunting of Leviathan: Seventeenth-Century Reactions to the Materialism and Moral Philosophy of Thomas Hobbes* (Cambridge: Cambridge University Press, 1962). For the sobriquet "Monster of Malmesbury," see *The True Effigies of the Monster of Malmesbury, or Thomas Hobbes in His Proper Colours* (London, 1680); Early English Books Online, ProQuest doc. 2264175032, accessed December 11, 2023. The anonymous author labels Hobbes a "Monster" in a poem that answers, stanza by stanza, Abraham Cowley's 1656 ode "To Mr. Hobs" (4).

[67] Thomas Hobbes, *De Homine* VII.1, in *The English Works of Thomas Hobbes*, ed. Sir William Molesworth, 11 vols (London: John Bohn, 1839–45), 4:31. For Milton's opposition to Hobbes on mechanism materialism and determinism, see my *Milton among the Philosophers*, 30–41, 206–43.

[68] Thomas Hobbes, *Leviathan*, ed. C. B. Macpherson (Harmondsworth: Penguin, 1968), 263 (Chapter 21).

Quadrangle; or *accidents of Bread in Cheese*; or *Immateriall Substances*; or of *A free Subject*; *A free-Will*; or any *Free*, but free from being hindred by opposition, I should not say he were in an Errour; but that his words were without meaning; that is to say, Absurd."[69] Despite his demurrals and his devoting of two of the four books of *Leviathan* to the "Christian Common-Wealth" and "the Kingdom of Darknesse," his sweeping dismissal of incorporeal substance struck many as leaving no place for God.

The threat to free will and to theism posed by Cartesian dualism was subtler. In *Meditations on First Philosophy*, Descartes offers several proofs of the existence of God, including in Meditation V a modified version of Anselm's Ontological Proof for the existence of God. In the *Discourse on the Method* and the *Meditations* Descartes offered a philosophical foundation for the existence of incorporeal substance alongside extended material substance, thus providing a place for an incorporeal God, angels, and immortal souls, and also for free will, which Descartes viewed as not merely real but "infinite."[70] Freedom of the will for Descartes was preserved by its location in *res cogitans*, exempt from the mechanism governing *res extensa*.

Despite Descartes' assertions of the existence of God and incorporeal substance and insistence on the faculty of free will, many came to fear the implications of his mechanist views on the operations of *res extensa*.[71] More laments to Robert Boyle "how prejudicial *Des Cartes's* mechanical pretensions are to the belief of a God."[72] And if a mechanist physics raised questions about the need for a creating God, it was also seen as leading, whatever Descartes' defense of free will, toward psychological determinism. For More and Cudworth, a realm of incorporeal substance restricted to angels and the human mind, set against an otherwise universal mechanism, seemed a limited and fragile haven for free will and for divine action in creating and maintaining the world. For leading English natural philosophers, as we will see below, the belief that physical matter works only mechanically seemed both a derogation of divine power and a reductionist begging of the question in explaining natural phenomena. English philosophers and experimentalists were skeptical of dualisms of both the Cartesian and Cambridge Platonist varieties. As was the case with the response to theories of causation by the interaction of passive matter, the objections were both philosophical and theological.

[69] *Leviathan*, 113 (Chapter 5).

[70] See *Meditations on First Philosophy*, Meditation IV.

[71] For the Cambridge Platonists' initial embracing and eventual repudiation of Descartes, see my *Milton among the Philosophers*, 21–30, 50–78.

[72] Letter of Henry More to Robert Boyle, December 4, 1671, in *The Conway Letters, The Correspondence of Anne, Viscountess Conway, Henry More and their Friends, 1642–1684*, ed. Marjorie Hope Nicolson and rev. Sarah Hutton (Oxford: Oxford University Press, 1992), 519; the year is not certain but probable.

An important philosophical objection was that two-substance dualism raised an unanswerable question: how could an incorporeal substance affect, move, or influence a corporeal body. Descartes' suggestion that they met in the human pineal gland specified the site of, but did not explain the interaction of, *res cogitans* and *res extensa*. The interaction is miniaturized and occulted rather than explained. The animist materialist Cavendish stated the difficulty succinctly: "material and immaterial are so quite opposite to each other, as it is impossible they should commix and work together, or act upon the other: nay, if they could, they would make but a confusion, being of contrary natures."[73] Cavendish believed in the infinite variety of an infinite nature created by an infinite God, leading to repeated assertions that there is more variety of existents and modes of causation in the world than we can ever know, but her infinites did not embrace the interaction of two distinct and separate substances. Malebranche's occasionalism, which posited that God is the efficient cause of all events, represents an attempt to deal with this objection to dualism; while the mind might, for example, will to raise the arm, God causes the arm to rise. This view was not taken up by English thinkers.

Newton objected to dualist, and particularly Cartesian, assertions of the existence of incorporeal substance on the principle that whatever is, must be in place: "Nor is the distinction between mind and body [*distinctio mentis a corpore*] in [Descartes'] philosophy intelligible, unless at the same time we say that mind has no extension at all, and so is not substantially present in any extension, that is, exists nowhere; which seems the same as denying the existence of mind [*quod perinde videtur ac si diceremus non esse*], or at least renders its union with body totally unintelligible, not to say impossible."[74] Newton's argument for the extension of mind, that is, for the extension of what some considered incorporeal, unextended substance, is theologically driven. Without extension, God could not be everywhere, Newton argues. And as God is omniscient as well as omnipresent, God as it were "senses" everything that happens everywhere. The extension of God is reflected in the extension of mind, as Newton continues: "if the distinction of substances between *thinking* and *extended* is legitimate and complete, God does not eminently contain extension within himself and therefore cannot create it; but God and extension will be two substances separately complete, absolute, and having the same significance."[75] Newton on the same page maintains that Cartesian dualism "offer[s] a path to Atheism."[76] Newton's position here

[73] Cavendish, *Observations*, "Argumental Discourse," 35.

[74] "De gravitatione," in Hall and Hall, *Unpublished Scientific Manuscripts of Isaac Newton*, 143 (Latin 109).

[75] "De gravitatione," 143 (Latin 109).

[76] Newton's conception of matter (in the sense of body) is, like Milton's, not Aristotle's *prima materia* or substrate without attributes. I differentiated Milton's from the scholastic and Aristotelian view earlier in this chapter on the basis of his depiction in *Paradise Lost* of the matter of chaos exhibiting qualities and of Raphael's description of the "things" of the "one first matter all." For his part, Newton

might seem indebted to that of his friend Henry More, who insisted on the extension of mind, but More also insisted on the distinction between mind and body, which Newton deems, as we have just seen, not "in philosophy intelligible."

The assumption that matter was by itself, in either its dualist or mechanist materialist forms, essentially inert also raised objections both philosophical and theological. The natural philosophical objection was that defining body as by definition inert and passive begs the question. It was premature and unwarranted to assume that the cause of movement could not be found in self-active matter. More clearly articulated just this objection aimed at his own position that incorporeal "Spirit of Nature" was necessary to move passive matter:

> *As for* the Spirit of Nature, *the greatest exceptions are, That I have introduced* an obscure Principle *for Ignorance and Sloth to take Sanctuary in, and to enervate or foreslack the useful endeavours of curious Wits, and hinder that expected progress that may be made in the* Mechanick Philosophy*; and this, to aggravate the crime, before a competent search be made what the* Mechanical Powers of Matter *can doe.*[77]

More's "curious Wits" include the group of experimentalist natural philosophers committed to the compatibility of natural philosophy and Christianity, such figures as Robert Boyle, Robert Hooke, and Matthew Hale, who came to be known as "Christian *virtuosi*," a name borrowed from Boyle's 1690 book *The Christian Virtuoso*. The experimentalists faulted More for needlessly multiplying substances before there was sufficient time and, along with time, improvements in instrumentation, to discover active principles in corporeal substance.[78] Hooke mocks More's conception of a "Hylarchick Spirit," another name for his "Spirit of Nature": "if all things be done by an Hylarchick Spirit, that is, I know not what, and to be found I know not when or where, and acts all things I know not how, what should should [sic] I trouble my self to enquire into that which is never to be

is explicit on the difference: "if we adopt the vulgar notion (or rather lack of it) of body, according to which there resides in bodies a certain unintelligible reality that they call substance, in which all the qualities of the bodies are inherent, this (apart from its unintelligibility) is exposed to the same inconveniences as the Cartesian view. For as it cannot be understood, it is impossible that its distinction from the substance of the mind should be understood. For the distinction drawn from substantial form or the attributes of substances is not enough: if bare substances do not have an essential difference, the same substantial forms or attributes can fit both, and render them by turns, if not at one and the same time, mind and body.... Further, they attribute no less reality in concept (though less in words) to this corporeal substance regarded as being without qualities and forms, than they do to the substance of God, abstracted from his attributes. They conceive of both, when considered simply, in the same way; or rather they do not conceive of them, but confound them in some common idea of an unintelligible reality. And hence it is not surprising that Atheists arise ascribing that to corporeal substances which solely belongs to the divine" ("De gravitatione," 143–44).

[77] More, *Immortality of the Soul*, Preface 11, in *A Collection*, vol. 2, 11. The first sentence is quoted in Henry, "Occult Qualities," 355.

[78] Henry, "Occult Qualities," 354–56.

understood, and is beyond the reach of my Faculties to comprehend?"[79] Hooke bridles at the assumption that motion and change must be directed by an incorporeal Spirit of Nature that by definition cannot be observed however powerful the microscope and air pump, or however sensitive the scales.

Boyle's and Hooke's opposition to More and other proponents of purely incorporeal means of causation derived, according to John Henry, from "the natural philosophers' belief...that active principles, for all their occult nature, were amenable to philosophical investigation while the Spirit of Nature was not."[80] These active principles, the experimentalists hoped, were only practically and temporarily occult; future advances in optical instruments might reveal what had hitherto been imperceptible. Boyle argued that in addition to the world of "objects proportioned to our unassisted sight" and a spiritual world of incorporeal angels and souls, there was a third world that natural philosophers could investigate, a "dioptrical world...which consists of all those creatures that lay concealed in former ages, from mortal eyes, and are not now discovered, without dioptrical glasses," that is, without lenses.[81] For these experimentalists, material but non-mechanical forces, such as those hypothesized by Newton, might ultimately be observable by instruments.

William Harvey among others challenged claims that matter could not on its own give rise to activity, or that matter prior to being organized in complex structures could not, as Aristotle argued, be animated.[82] In observing under a microscope the stages of conception and embryos, Harvey had detected the irritability of tissue fibers before the formation of embryo, and he concluded from that fact that life was already present in simple, unorganized matter.[83] In *Anatomical Exercises on the Generation of Living Creatures*, Harvey observes,

> There appears...from the beginning, *Punctum Rubrum saliens, Vesicula pulsans, Fibraeque inde deductae* [twitching red-colored point, a pulsating little bladder, with lines stretching out from it], the Red capering Point, the Panting Vesicle, and the Fibers deduced thence, which contain *blood*. And so farre as I could possibly discern by accurate Inspection, the *Blood* is formed before the *Punctum*

[79] Robert Hooke, *Lampas, or, Descriptions of Some Mechanical Improvements of Lamps & Waterpoises together with Some Other Physical and Mechanical Discoveries* (London, 1677), 34; The Making of the Modern World, Gale doc. GALEIU0104138027, accessed February 23, 2024.

[80] Henry, "Occult Qualities," 355.

[81] Boyle, *The Christian Virtuoso*, quoted in Catherine Wilson, *Epicureanism at the Origins of Modernity* (Oxford: Clarendon, 2008), 230. For the balance of conviction and doubt in Boyle's picture of the Christian Virtuoso, see Wilson, *Epicureanism at the Origins of Modernity*, 224–51.

[82] See Charlotte Nicholls, "Body Out of Spirit: Medical Science and the Creation of Living Soul in *Paradise Lost*," *Milton Studies* 57 (2016): 134.

[83] Ann Thomson, *Bodies of Thought: Science, Religion, and the Soul in the Early Enlightenment* (Oxford: Oxford University Press, 2008), 69.

> *saliens*, and is endowed with *vital heat*, before it moves by *pulse*; and in it, and from it doth the *pulsation* begin.[84]

For Harvey, the matter of the blood can possess vital heat and hence life before a complex organization of matter. Life, he concluded, is a product neither of incorporeal spirit nor of organized matter; instead, vital matter gives rise to organized bodies.

The denial of vital, active power to matter was, according to the experimentalists, not only premature but, in theological terms, a mistaken circumscription of divine omnipotence. Sensitive to charges that locating the sources of life and motion in matter resembled what were thought to be the atheistic opinions of Epicurus and Lucretius, Boyle and Hooke based their confidence that active principles could be found in their belief that an omnipotent God can do anything. If the rationalist Cambridge Platonists insisted that corporeal matter by its essence must be inactive, voluntarist natural philosophers responded that there was no such limitation on the power of God, who could create corporeal active principles.[85] Newton, as we shall see, would echo these *virtuosi* in emphasizing God's power as warrant for the possibility of active and living matter.

While charges of atheism were leveled against materialists and vitalists in the seventeenth century, Ann Thomson points to the committed theism of vitalist natural philosophers.[86] Some argued that the incorporeality and natural immortality of the soul had no biblical warrant, adding that, if the soul is immortal, it is by the ordination of God, not by its essence as an incorporeal substance. John Locke famously suggested that it is "not much more remote from our comprehension to conceive that God can, if he pleases, superadd to matter a faculty of thinking, than that he should superadd to it another substance [i.e., incorporeal spirit] with a faculty of thinking."[87] Some argued that God did superadd to matter the faculty of thinking, whether in the form of the essential perceptive ability of all matter, as in the case of the iatrochemist Francis Glisson, discussed in the next chapter, or only in the form of organized bodies of matter such as the brain.

The voluntarist divine Richard Baxter repudiated the assumption that matter is essentially inert as a slur on God's omnipotence: "I confess I am too dull to be sure that God cannot endue *matter* itself with the formal Virtue of Perception.... Nor do I believe that it giveth a man any more cause to doubt... of the *Existence of God*,

[84] William Harvey, *Anatomical Exercitations, Concerning the Generation of Living Creatures* (London, 1653), 274, from the chapter "Of the Order of Generation, and, first, of the first Genital Particle"; Early English Books Online, ProQuest doc. 2240960739, accessed August 7, 2025.

[85] Henry, "Occult Qualities," 342.

[86] Thomson, *Bodies of Thought*, 76–77; I draw in this paragraph from Thomson.

[87] Locke, *An Essay Concerning Human Understanding*, ed. Pauline Phemister (Oxford: Oxford University Press, 2008), 345 (IV.3.vi). For the philosophical context of Locke's speculation, see John Yolton, *Thinking Matter: Materialism in Eighteenth-Century Britain* (Minneapolis: University of Minnesota Press, 1983).

or the *Immortality of the Soul*, than your Opinion that saith, *God cannot do this*."[88] Boyle exemplifies Baxter's point, as he maintained at the same time both the activity of matter on the basis of reason and experiment and the immortality of the soul based on divine revelation.[89] Boyle insists on the authority of Scripture that human souls are incorporeal and immortal,[90] while at the same time admitting active principles in body. In a letter to Milton's friend Samuel Hartlib composed before 1660 and published posthumously in 1692, Boyle refers to material spirits "of an Aerious, Ethereal, Luminous Production and Composition" present in "all mix'd Bodies," which are kindled into life by Spirits descending from the sun:

> As the Sun-shining on the rest of the Planets doth not, as we said, only barely illuminate their Bodies; but besides this, through the Power, Virtue and Activity it hath, doth also raise, excite, awaken and stir up the several Properties and Dispositions that are in those several and respective Bodies, whereby they are more lively and effectually brought forth upon us; so we are to suppose it is in reference to this our Planet, which is the Earth, which is not only enlightned, warmed, cherished and fructified by the Power, Virtue and Influence of the Sun; but hath its proper Magnetical, Planetary Virtue, also fermented, stirred, agitated and awakened in it.[91]

These Spirits comprise a stream of "effluviums of the air" that possess vital energy and awake dormant spirits in mixed bodies, "stir[ing], agitat[ing] and awaken[ing]" them without the intervention or direction of incorporeal spirits. Like Locke, Boyle thought that an omnipotent God could endow matter or body with the power to think, but he believed that God did not do so.[92] Others, including Milton, thought that God not only could but also did endow matter with the power to think, and Cavendish would go so far as to argue that all bodies think in at least rudimentary fashion.

Cavendish, like the experimentalists discussed above, pointed to divine omnipotence to support her argument for living, active matter, while at the same time questioning how something immaterial could move something material. "If that immaterial spirit" hypothesized by More and others, she asked, "have

[88] Baxter, *Of the Nature of Spirits, Especially Mans Soul* (London, 1682), 28–29, quoted in Thomson, *Bodies of Thought*, 76.

[89] Thomson, *Bodies of Thought*, 75.

[90] For Boyle on the immortal soul and the dualism of soul and body, see his *The Excellency of Theology Compar'd with Natural Philosophy (as both are Objects of Men's Study) Discours'd of in a Letter to a Friend. To which are annex'd some Occasional Thoughts about the Excellency and Grounds of the Mechanical Hypothesis* (London, 1674), 25; Early English Books Online, ProQuest doc. 2264186062, accessed July 9, 2025.

[91] Robert Boyle, *The General History of the Air* (London, 1692), 73–75; Early English Books Online, ProQuest doc. 2240859706, accessed December 11, 2023. For the dating of the letter and for Boyle's openness to active principles, see Henry, "Occult Qualities," 344.

[92] See Thomson, *Bodies of Thought*, 78–79.

self-motion, why may not a natural corporeal being, have the like? they being both creatures of God, who can as well grant self-motion to a corporeal, as to an incorporeal being: nay, I am not able to comprehend how motion can be attributed to a spirit; I mean, natural motion, which is only a propriety of body, or of a corporeal being."[93] Cavendish presents the limit case for active, knowing matter in seventeenth-century England. Complex bodies for her are organized by the parcels of matter constituting them, which can do so only because they can perceive, know, and organize the larger bodies to which they belong.[94]

If Milton viewed life and self-motion as qualities of matter rather than requiring the presence and causative influence of incorporeal substances, he has much in common with the leading figures of seventeenth-century English experimentalism. While there was a spectrum of opinion on whether self-active matter could be the basis of thought and on whether the human soul might be corporeal or incorporeal, Locke's famous speculation that God might "superadd to matter the faculty of thinking" was by 1689, when the *Essay* was published, more conventional than not.

The Theological Springs of Milton's and Newton's Vitalisms

Milton and Newton shared the motivations as well as the conclusions of contemporaries who accepted that matter could be self-active. Newton's commitment to defending divine omnipotence was a strong motivation. He shared with the experimentalist *virtuosi* the experience of identifying mechanical causation and formulating mechanical laws, and at the same time seeing evidence that not all phenomena, including the preservation of motion in the universe, could be explained mechanically. With others, Newton turned to divine omnipotence to support arguments that there are two kinds of causation in the universe. One is mechanical, involving the movement of bodies of passive matter by collision or interaction with other bodies of passive matter. The other is based on active principles or self-active matter. Even Cavendish, who argued that *all* matter is living, perceptive, knowing, and self-active, acknowledged two kinds of causation in the created universe. She writes that in arguing that "natural actions are free…, and not constrained; my meaning is not, as if there was no pressing or driving of parts at all in nature, but only that they are not the universal or principle actions of nature's body, as it is the opinion of some philosophers, who think there is no other motion in nature, but by pressure of parts upon parts."[95] Conway agreed on this point: "For truly in nature there are many operations that are far more than

[93] Cavendish, *Observations*, I.27 (111–12). [94] Cavendish, *Observations*, I.27 (113).
[95] Cavendish, *Observations*, "To the Reader" (19).

merely mechanical. Nature is not simply an organic body like a clock, which has no vital principle of motion in it; but it is a living body which has life and perception, which are much more exalted than a mere mechanism or a mechanical motion."[96] There is a strong strain of voluntarism in English seventeenth-century natural philosophy, which rebelled against rationalist *a priori* principles, such as the Cambridge Platonist insistence that matter is essentially inert, passive, and inactive.

While Milton, like all Christian thinkers, asserted divine omnipotence, it was his commitment to freedom of the will that informed and motivated his vision of living matter. This imperative is visible in his works from the beginning to the end of his career. It is famously at the center of *Areopagitica*: "when God gave [Adam] reason, he gave him freedom to choose, for reason is but choosing; he had bin else a meer artificiall *Adam*, such an *Adam* as he is in the motions" (*CPW* 2:527). The theodicy of *Paradise Lost*, which aims to "justify the ways of God to men" (*PL* 1.26), depends in his eyes on the narrative embodiment of human free will. Cavendish and Conway also address the relation of vitalism to free will that I argue is important for Milton, as they had addressed the connection of vitalism and divine omnipotence that I argue is important to Newton.

While only God in Cavendish's "holistic materialism" has absolute freedom, self-moving matter has a large range of freedom:

> nature is neither absolutely necessitated, nor has an absolute free will: for, she is so much necessitated, that she depends upon the all-powerful God, and cannot work beyond herself, or beyond her own nature; and yet hath so much liberty, that in her particulars she works as she pleaseth, and as God has given her power; but she being wise, acts according to her infinite natural wisdom, which is the cause of her orderly government in all particular productions.[97]

While Cavendish reserves absolute freedom for the "all-powerful God," there seem to be few practical limits on the free working of nature and her wisdom. The constraints on nature's freedom are more apparent than real, for Cavendish goes on to argue that "natural self-motions" are not merely undetermined by "parts [that] do drive and press upon each other" (i.e., by mechanical causation), but are "free and voluntary."[98]

> no particular parts are bound to certain particular actions, no more than nature herself, which is self-moving matter; for as nature is full of variety of motions or actions, so are her parts; or else she could not be said self-moving, if she were bound to certain actions, and had not liberty to move as she pleases: for though

[96] Conway, *Principles*, 64 (Chapter 9).
[97] Cavendish, *Observations*, I.27 (109).
[98] Cavendish, *Observations*, I.31 (127).

> God, the author of nature, has ordered her so that she cannot work beyond her own nature, that is, beyond matter; yet has she freedom to move as she will.[99]

Given that in Cavendish's philosophy nothing is "beyond matter," the restriction on the freedom of the parcels of matter turns out to be vanishingly small. God is the one immaterial in or beyond Cavendish's universe.[100]

The same pattern of an apparent placing of limits on the freedom of matter being eclipsed by assertion of freedom characterizes Cavendish's treatment of rational human freedom. In the *Observations*, she affects a kind of agnosticism on free will and predestination: "It is true, many persons are much troubled concerning free will and predestination, complaining, that the Christian church is so divided about this article, as they will never agree in one united belief concerning this point; which is the cause of the trouble of so many consciences, nay, in some even to despair."[101] While bracketing this question and leaving it to the theologians, she solves the existential dilemma by assuring readers that if they but love God, pray, not harm others, live virtuously and piously, and follow the teaching of the (admittedly divided) Christian church, they need not despair.

A decade earlier, in the 1656 *Natures Picture*, Cavendish apparently denies free will to human beings, in the voice of an alter-ego, the She-Anchoret: "for, said she, if *Jove* had given Men Free-Will, he had given the use of one of his Attributes to Man, as free Power; which, said she, *Jove* cannot do, for that were to lessen himself, To let any Creature have free power to do what he will."[102] This a position to warm predestinationists' hearts. Nevertheless, the She-Anchoret, speaking for Cavendish, goes on to ascribe to human beings the kind of freedom familiar in Milton, and with a similar rationale. If human beings were not free to choose their actions, if God were to command actions we are not free to choose and beliefs we are not free to hold, this would be "as it were, to force him [i.e., man] to disobey him: or to think *Jove* suffers Man to do evil, when he could prevent it; or to think *Jove* permits Man to provoke his Justice, or to damn Man, when it is in *Jove*'s power to save him, were to think *Jove* unjust and cruel."[103] The She-Anchoret here agrees with the Milton of *De Doctrina Christiana*: if God were to command us to do what we are not free to choose to do, God would be the author of evil.[104]

[99] Cavendish, *Observations*, I.35.13 (139).

[100] The Blazing World of Cavendish's work of the same name, published in the same volume with the *Observations*, and the other worlds invented inside *The Blazing World* turn out to be immaterial, an exception owing to those worlds' fictionality; see my "Imagining Materialism in *Paradise Lost* and *the Blazing World*."

[101] Cavendish, *Observations*, II.12 (217–18). See also Sarasohn's discussion (116) of a similar position in the 1664 *Philosophical Letters*.

[102] Margaret Cavendish, *Natures Picture Drawn by Fancies Pencil* (London, 1656), 613; Early English Books Online, ProQuest doc. 2240924768, accessed December 11, 2023. The passage is quoted in Sarasohn (90), who provides the EEBO image number rather than the page number.

[103] *Natures Picture*, 613–14.

[104] Milton, echoing the anti-predestinarian Arminius, writes that if God predestines individuals to reprobation, "he also decreed the means without which he cannot carry out that decree of his; but the means is sin alone" (*De Doctrina Christiana*, I.4, *OCW* 8.1.99–101).

Cavendish's apparent denial of free will to human beings is in the service of her characteristic refusal to elevate human nature above nature in general. A God who would give greater freedom of the will to human beings than to other creatures—and even to what one might think to be inanimate objects—would be showing partiality. Cavendish ascribes knowledge of self and God not only to what most think inanimate objects (such as stones) but even to the "inanimate parts of matter" found, along with "animate parts," in every parcel of matter. She is polemical against the hubris of human beings, who see themselves as the pinnacle of creation, noting that all creatures perceive and know after their own manner, and that so-called lower animals know things that human beings do not. She argues that "we cannot say a bird is a more perfect figure than a beast, or a beast a more perfect figure than a fish, or worms; neither can we say that man is a more perfect figure than any of the rest of the animals: the like of vegetables, minerals and elements."[105] Cavendish's philosophy is more radically leveling than Milton's, who has Raphael in *Paradise Lost* compress the distance between ontological levels ("time may come when men / With angels may participate"; "what if Earth / Be but the shadow of Heav'n, and things therein / Each to other like, more than on Earth is thought" [5.493–94, 574–76]). For Cavendish, the leveling may owe something to rebellion against the universal assumption of male superiority, against which she clearly bridles. The denial of an incorporeal soul is an essential part of this leveling:

> why might not infinite matter be decreed to move of itself as well as a spirit, or the immaterial soul? But I perceive man has a great spleen against self-moving corporeal nature, although himself is part of her, and the reason is his ambition; for he would fain be supreme and above all other creatures, as more towards a divine nature: he would be a God, if arguments could make him such, at least God-like, as is evident by his fall, which came merely from an ambitious mind of being like God.[106]

[105] Cavendish, *Observations*, II.6 (204). Shortly afterward Cavendish elaborates on animals' knowledge: "Some philosophical writers discourse much concerning the knowledge of man, and the ignorance of all other creatures.... I believe other creatures have as much knowledge as man...; but their knowledges being different, by reason of their different natures and figures, it causes an ignorance of each other's knowledge; nay, the knowledge of other creatures, many times gives information to man: As for example; the Egyptians are informed how high the river Nile will rise by the crocodile's building her nest higher or lower; which shows, that those creatures foresee or foreknow more than man can do: Also, many birds foreknow the rising of a tempest, and shelter themselves before it comes; the like examples might be given of several other sorts of animals, whose knowledge proceeds either from some sensitive perceptions, or from rational observations, or from both; and if there be such a difference in the rational and sensitive knowledge of one kind of creatures, to wit, animals, much more in all other kinds, as vegetables, minerals, elements, and so in all nature's works: Wherefore he that will say, there is no knowledge but in man, at least in animal kind; doth, in my opinion, say more than ever he will be able to prove" (Cavendish, *Observations*, II.12 [218]).

[106] Cavendish, *Observations*, II.7 (209).

Men claim incorporeal souls, Cavendish suggests, out of a vainglorious desire to separate themselves from nature and attach themselves to the supernatural, to claim the absolute free will of God as opposed to being submerged in the soulless mechanism of the rest of creation. This hubris is for Cavendish a particularly masculine trait. *Man*, *he*, *him*, and *his* in this passage can be read as referring generically to the human being, but Cavendish's frequent bridling at the exclusion of women from universities and learned societies, as well as her experience of having her philosophy ignored by her male contemporaries, suggests that the animus here is directed toward men in particular. While theoretically reserving the greatest degree of freedom for God, Cavendish envisions a world of free, self-moving animate matter, which underwrites the freedom of, but not only of, the human will.

For Conway, the foundation of freedom is in matter's "life and perception, which are much more exalted than a mere mechanism or a mechanical motion."[107] They make possible "these more excellent attributes…: spirit or life and light, by which I mean the capacity for every kind of feeling, perception, or knowledge, even love, all power and virtue, joy and fruition, which the noblest creature have or can have, even the vilest and most contemptible," including "indifference of the will" or freedom.[108] Like Milton and Cavendish, Conway moves from the freedom of matter to the freedom of the will to theodicy: "if [the creature] moves against [God's] will…, then it sins and is punished for it. But God is not the cause of sin,…because when a creature sins, it abuses that power which God granted it. Thus the creature is culpable and God is entirely free from every spot of sin."[109] The progression in Conway is seamless. Opposition to Hobbes' mechanism and a philosophy of the life of matter underwrite a defense of both freedom of the will and the justice of the ways of God to men. Cavendish and Conway make explicit the intimate connection that I have argued in the past connects Milton's passionate defense of the freedom of the will and his conception of the life of matter.

Milton's and Newton's Living Universes

Milton's and Newton's theologically driven understandings of the life of matter help to account for similarities between their ways of writing about creation, even as their perspectives as natural philosopher/theologian and poet/theologian distinguish them. Responsible, unlike Milton, to flesh out a natural philosophy and square it with empirical observation, Newton's restless mind oscillated on some matters—he posited a subtle material aether in space, rejected it, posited it again, and rejected it again. From the time of his initial alchemical explorations in

[107] Conway, *Principles*, 64 (Chapter 9).
[108] Conway, *Principles*, 66, 15 (Chapters 3 and 9).
[109] Conway, *Principles*, 70 (Chapter 9).

the 1670s, however, he worked toward an understanding and articulation of the life of matter. In his Draft Queries to his *Opticks*, he writes, as we have seen, "We cannot say that all Nature is not alive."[110] This vision of universal life is shared by Milton and Newton, not only in a general sense but also in particulars. In addition to asserting the life of all matter, including bodies thought by most to be inanimate, Milton and Newton shared a sense of the cosmos as a living organism. The universe of *Paradise Lost* is alive and even endowed with will. In the speech on "one first matter" we can observe a process of digestion that links in an unbroken chain the roots of plants and human reason. Through digestion, Raphael argues, the "corporeal to incorporeal turn[s]" (*PL* 5.413). And the poem suggests that the signs of life extend beyond and below what we take to be the animated and organic. The devil Belial foresees that the rebellion in Heaven will trigger something like a universal immune system:

> th'Ethereal mould
> Incapable of stain would soon expel
> Her mischief, and purge off the baser fire
> Victorious. (2.139–42)

Blurring further the distinction between the animate and the inanimate, Milton, through Raphael again, treats planets and elements as living and organic:

> For know, whatever was created, needs
> To be sustained and fed; of elements
> The grosser feeds the purer, earth the sea,
> Earth and the sea feed air, the air those fires
> Ethereal, and as lowest first the moon;
> Whence in her visage round those spots, unpurged
> Vapors not yet into her substance turned.
> Nor doth the moon no nourishment exhale
> From her moist continent to higher orbs.
> The Sun that light imparts to all, receives
> From all his alimental recompense
> In humid exhalations, and at even
> Sups with the ocean. (*PL* 5.414–26)

Paradise Lost is replete with the exhalations and sublimations of an animate earth.[111] Newton for his part remarks in his manuscript on "Nature's obvious laws" that it is "very agreeable to natures proceedings to make a circulation of all

[110] Query 23 in the "Draft version of Queries to the *Optics*," fol 620^{v}, accessed August 7, 2025.
[111] See *Paradise Lost* 3.585–610, 5.185, 5.642, 7.255, 10.694, 11.741.

things. Thus this Earth resembles a great animall or rather inanimate vegetable, draws in aethereall breath for its dayly refreshment & vitall ferment & transpires again with gross exhalations."[112] The earth transpires vapors that feed the sun, and the sun breeds metals, which grow in the bowels of the earth. In a manuscript letter read at a meeting of The Royal Society, to which we shall return in the next chapter, Newton writes that a "subtil" fluid arises from the earth and ascends to the atmosphere, and "perhaps may the sun imbibe this spirit copiously, to conserve his shining." One "may also suppose," he adds, "that this spirit affords or carries with it thither the solary fewel and material principle of light."[113] What might seem a merely fanciful image in Milton turns out to have a solid natural philosophical twin.[114] As Charles Webster demonstrates, the organic picture of the cosmos visible in Newton had a distinguished pedigree among earlier thinkers, including Kepler.[115]

What lessons should we draw from the surprising set of parallels between Milton and Newton? Why should one care about these resemblances? To say that they are interesting because they link the thought of towering contemporaries who seem in so many ways entirely unlike may not be sufficient, though I believe that to be part of the answer. A more compelling answer to my mind is that the set of resemblances draws our attention to the architecture of each figure's thought and to the rules governing intellectual discourse in seventeenth-century England. It was a world in which one's conception of God informed one's thinking in natural philosophy as well as theology. It was a condition of thought for most; for others (perhaps, e.g., Hobbes) it was at the least an imperative one had to acknowledge. For Milton and for Newton, the world was the way it was not because God, like the rationalist God of the Cambridge Platonists, had to create it in precisely the manner that he did. For Milton God created a world of living matter because in his goodness and justice he provided for the freedom of creatures; for Newton an omnipotent God could have created any world and any variety of causes in the world, and as a result he was not tied to solely mechanist causation or solely animist causation. To see the ways in which Newton, the banisher of poetry, articulates a world very like the world of *Paradise Lost*, is to free oneself from the illusion that what is important in Newton is what in hindsight

[112] "Of natures obvious laws," folio 3^{v}, http://webapp1.dlib.indiana.edu/newton/mss/norm/ALCH00081, accessed August 11, 2018.

[113] "An Hypothesis Explaining the Properties of Light," *The History of the Royal Society*, vol. 3 (London: A. Millar, 1757), 251. The source is the December 7, 1675 letter to Milton's friend Oldenburg, discussed above (*Correspondence*, I:366).

[114] Denis Saurat finds an analogue for this image in Robert Fludd's *De macrocosmi historia* I.i.6 and 7 (137 and 139), in Denis Saurat, *Milton, Man and Thinker* (New York: Dial Press, 1925), 305–6.

[115] According to Webster, "The earth and other planetary bodies were not merely regarded as sharing some diffuse force characterized as the *anima mundi*, but were also thought to possess a complete organic constitution analogous to the human body. The planetary body was revitalized by the circulation of its physiological fluids," in Charles Webster, *From Paracelsus to Newton: Magic and the Making of Modern Science* (Cambridge: Cambridge University Press, 1982), 89–92.

has seemed to be the creation of classical mechanics. To see these similarities is to obtain a purchase on the strangeness of their world, which can give us a new perspective on our own.

At the beginning of the seventeenth century, John Donne ruefully observes that the "new philosophy calls all in doubt, / The element of fire is quite put out; / The sun is lost, and th'earth, and no man's wit / Can well direct him where to looke for it."[116] But for Milton and Newton and for many of their scientific contemporaries, the element of fire and the sun burned still with life-giving force. Milton's and Newton's thinking on the cosmos and on the relation of body and spirit participated in a discourse that assumed the animation of the world and the availability of divine signs in the heavens and on Earth. For natural philosophers from Paracelsus and Kepler to Newton, as Webster has argued, "Acceptance as a normal part of physics of such devices as an animistic interpretation of the earth and celestial bodies, and the guiding concept of the *anima mundi*, rendered it easy to posit connections between the physical and organic and psychic worlds."[117] The providential plan of God was written into the cosmos, into the microcosm, and into history. Theologians tracing biblical chronology and signs of the millennium in the Book of Revelation followed the work of astronomers and physicists and vice versa, and Newton carried on both inquiries simultaneously.

To understand Newton, one needs to acknowledge that he was driven by an imperative to find evidence of God's omnipotence in Scripture, in the heavens, in the vegetation of metals, in biblical prophecy, in human history, in short in any and all of his apparently disparate intellectual pursuits. Descriptions of an animate universe, of gravity, of biblical chronology, of prophecies of the millennium—all elucidate the power and glory of God. Milton also looked to an animate universe and to prophecy and history to find evidence of what for him is central and fundamental—freedom that characterizes not only human will but also God's and even what we usually think of as unreasoning, non-sentient, and inorganic.

Newton's emphasis on divine omnipotence over human freedom is vividly illustrated by the omission of any mention of free will in his notes on Cudworth's *True Intellectual System*, a work centrally devoted to the defense of free will. Newton to my knowledge mentions free will only once, lumping it with predestination and "the state of the dead between death & the resurrection" as a matter of knowledge which is not necessary for salvation.[118] The difference in emphasis

[116] John Donne, "An Anatomy of the World: The First Anniversary," ll. 205–8, in *Donne: Poetical Works*, ed. Sir Herbert Grierson (Oxford: Oxford University Press, 1933), 213–14.

[117] Webster, *From Paracelsus to Newton*, 30.

[118] See Newton's manuscript "Out of Cudworth" in the William Andrews Clark Library, shelfmark fN563Z; they are transcribed in an appendix to James E. Force and Richard H. Popkin, ed., *Essays on the Context, Nature, and Influence of Isaac Newton's Theology* (Dordrecht: Kluwer, 1990), 207–13. For Newton's cursory and non-committal comment on free will, see his manuscript "Draft on the history of the Church," Yahuda ms. 15, 134^{r-v}, http://www.newtonproject.ox.ac.uk/view/texts/normalized/THEM00237, accessed August 11, 2018.

between the thinkers is on display when Milton in *Paradise Lost* and Newton in the General Scholium of the *Principia Mathematica* distance their Gods from pagan notions of Fate. Milton's God insists that in creating He "is free / To act or not, necessity and chance / Approach not me, and what I will is fate" (*PL* 7.168–73). Newton writes that one can know God "only by his... wisest and best construction of things and their final causes, and we admire him because of his perfections; but we venerate and worship him because of his dominion.... [A] god without dominion, providence, and final causes is nothing other than fate and nature."[119] Each emphasizes God's unfettered activity, but where Milton stresses the freedom of God's will, Newton stresses God's unlimited power.

Nevertheless, Milton and Newton share, as I have argued, an antipathy to dualism and mere mechanism as inimical to a proper conception of the relation of God and creation. This shared antipathy is visible in what I have argued are remarkably similar pictures of an organic, living universe. It is also visible in shared deviations from orthodox Christian teaching, notably mortalism, to which I will return in Chapter 4. The shared mortalism of Milton and Newton, however much it is based in scripture, reflected their understandings of the relation of spirit and body.[120] Life is not the peculiar characteristic of a separate, incorporeal substance; instead it inheres in rarefied corporeal spirits. There is no ontological divide, with immortal incorporeality on one side and mortal corporeality on the other.[121] For Milton, as we have seen, "the idea of man's spirit being separated from his body... is not only not found anywhere in holy scripture, but also plainly clashes with nature and reason" (*De Doctrina Christiana*, I.7; *OCW* 8:303–5).

Milton and Newton offer intriguingly complementary pictures of the proper location of the tenuous spirits. Newton proposes that the active, tenuously corporal spirits that vivify grosser bodies descended from the heavens; the tails of comets might supply the spirit needed for life on earth: "I suspect that that spirit which is the smallest but most subtle and most excellent part of our air, and which is required for the life of all things, comes chiefly from comets."[122] In the *De Doctrina*

[119] "General Scholium," *The Principia*, trans. Cohen and Whitman, 942.

[120] Stephen Snobelen collects passages containing Newton's mortalism, to be discussed in the Chapter 5, in "Isaac Newton, Heretic: The Strategies of a Nicodemite," *British Journal for the History of Science* 32, no. 4 (1999): 387, n. 44; key passages are in "Paradoxical Questions concerning the morals & actions of Athanasius & his followers," William Andrew Clark Memorial Library, shelfmark N563M3 P222, the Clark MS, ff. 54r–55r, https://www.newtonproject.ox.ac.uk/view/texts/diplomatic/THEM00010, accessed August 18, 2025, and in "Of the Woman & Two-horned Beast," Yahuda MS 7.2e, "ye resurrection from ye dead is called living again & therefore between death & the resurrection men do not live," f. 4v; https://www.newtonproject.ox.ac.uk/view/texts/diplomatic/THEM00424, accessed August 7, 2025.

[121] Newton blames the erroneous belief in the immortality of the soul on his detested Athanasius, whom he also blames for the trinitarianism that both he and Milton reject. See the ms. "Paradoxical Questions," fol. 54r. See Stephen Snobelen, "Lust, Pride and Ambition: Isaac Newton and the Devil," in *Newton and Newtonianism: New Studies*, ed. James E. Force and Sarah Hutton (Dordrecht: Kluwer, 2004), 164.

[122] Newton, *Principia*, Book III, Prop. 41 (926).

Christiana, Milton speculates that at death the subtle spirits disperse and rise to their source in the heavens, significantly mirroring and closing the circuit of Newton's speculation:

> God is said [in Eccles. 12:7] *both to have given and to take back to himself* the spirit of absolutely all animate beings when the body returns to dust, Job 34:14–15. . . . How much more accurately [than my opponents] has **Euripides**, perhaps unwittingly, interpreted this passage in his **Suppliant Women** [532–34]:
>
> And whence each part came into the light
> There it returns: spirit back to air,
> The body back into earth . . .
>
> That is, each dissolved part returns into its own origins, into its own elements; which is also confirmed by Ezekiel, ch. 37:9: *come from the four winds, O spirit*: certainly, then, man's spirit had gone away to that place whence it returned. (*DDC* I.13; *OCW* 8:455)

The tenuous spirits return to the skies while the grosser matter falls to the earth, just as in Newton the subtle active spirits come from the heavens. Milton's claim that the spirits will be dispersed (and eventually regathered) as the "tiniest little dust-specks of their bodies [*minutissimos corporum pulvisculos*]" (OCW 8:454–55) can be related to Newton's corpuscularian or atomist account of the substrate of bodies.[123] Linking cosmos and Earth, celestial and terrestrial animate bodies, active material spirits inform Milton's and Newton's understandings of creation.

Milton and Newton: Matter, God, and Modernity. Stephen M. Fallon, Oxford University Press.
 DOI: 10.1093/9780198988939.003.0002

[123] For Newton's atomism, see Query 31 of the *Opticks* and his "De natura acidorum," Cambridge University Library MS, shelfmark Add. 9597/2/18/81; this manuscript appears in translation in *Newton: A Norton Critical Edition*, ed. I. Bernard Cohen and Richard S. Westfall (New York: Norton, 1995), 312–13. The atomism of the universe of *Paradise Lost* is evident in the description of chaos (2.890–914).

2
Vitalist Alchemy

In the preceding chapter I explored parallels in the monist and vitalist thinking of Milton, Newton, and contemporary natural philosophers and metaphysicians. I turn now to one branch of increasingly vitalist natural philosophy, sometimes misunderstood in retrospect as a haven for mystics and charlatans: alchemy, whether chrysopoeia, the theory and practice of the transformation of metals, or iatrochemistry, the application of chemical principles to medicine. Acquaintance with early modern alchemical thinking can help us to better grasp Milton's animist materialism and Newton's efforts to understand the active principles that accounted for changes not reconcilable with merely mechanical causation.

Alchemy as Natural Philosophy, Not Mysticism

Alchemy and chemistry in the ancient, medieval, and early modern worlds were not separate enterprises. The English word "alchemy" is derived, through the Anglo-Norman *alkemye* and Middle French *alkemie*, from the Arabic *al-kīmiyā*, or "the chemia," itself derived from the Hellenistic Greek *chemia/χυμεία*. The word "chemistry," as might be obvious, shares with alchemy this etymology, and the sciences, despite anachronistic perspectives on alchemy, are all but indistinguishable. Both represent attempts to discover and understand the basic constituents of matter, and they share the ambition of manipulating and transforming matter. The ambition to turn base metals to gold was founded not on occult and mystical beliefs but on the confidence that the chemical makeup of things could be discovered and manipulated. The early modern alchemist can be said to be following the direction of Francis Bacon, who insisted that "nature to be commanded must be obeyed," meaning that one must understand the nature and causes of physical phenomena in order to be able to manipulate those causes to achieve one's ends.

Historians of early modern science have in recent decades established the scientific as opposed to mystical or occult nature of medieval and early modern alchemy. The appearance of occultism in early alchemical texts owes more to the intentional obscuring of potentially dangerous knowledge (think of the effect on economic systems of the publication of a method of turning lead into gold) than to any putative mystical element in alchemical teaching. As Lawrence Principe has shown, the idea that early alchemy is a form of magic rather than science or natural philosophy arose in the second half of the nineteenth century, owing to Mary Anne Atwood (1817–1910), who, influenced by the theories of Anton Mesmer, recast alchemy as involving "a self-transformative psychic practice" and promoted the idea of the alchemist's "self-purification and...exaltation to a 'higher plane of existence.'"[1] Where early modern alchemists such as Sendivogius, whom we will soon meet, and Newton worked by experimentation with metals, ores, salts, and acids, Atwood divided alchemy into the mere "'exoteric' (the language of chemical operations) and 'esoteric' (the hidden spiritual transformation)."[2] Principe shows that Atwood's unfortunately influential conceptualization of alchemy, while indicative of a climate of ideas in nineteenth-century New England, "[a]s a historical interpretation or description of pre-nineteenth-century alchemy...is simply wrong."[3]

Newton was interested in alchemy not for the promise of personal spiritual transformation but rather for the promise of understanding the activity in the natural world of God's omnipotent creative power, and specifically the nature and operation of active principles or active spirits. There was in the early modern world a tradition of alchemical *metaphors* for spiritual refinement (as gold is a refinement of baser metals, so the glorified body of the regenerate is a refinement of a baser mortal body), prompted in part by the desire of medieval European writers to Christianize a pursuit with Arabic Islamic roots.[4] But Newton's focus, characteristically, was on God's omnipotent power in the natural world rather than on the mechanisms of salvation. For him, as we will see in Chapter 3, subscription to a small number of basic doctrinal tenets was sufficient for that purpose. Not merely interested in alchemy, Newton was a working alchemist. In addition to voluminous notes on alchemical books, and transcriptions of and notes on numerous alchemical manuscripts, we have from his hand notes on his extensive alchemical experiments, records of his purchases of alchemical equipment, and even his own designs for alchemical furnaces. It was to alchemy that Newton turned in his attempts to discover "natures active laws," laws that would penetrate more deeply into God's creation than the "passive laws" of seventeenth-century mechanist natural philosophy.

The case for Milton's investment in alchemy requires more argument, for he was a poet rather than an experimental natural philosopher.[5] How are we to take the alchemical imagery that appears frequently and prominently in *Paradise Lost*,

[1] Lawrence Principe, *The Secrets of Alchemy* (Chicago: University of Chicago Press, 2013), 96; Principe refers to, and in the last phrase cites from, Mary Anne Atwood's *A Suggestive Inquiry into the Hermetic Mystery: With a Dissertation on the More Celebrated of the Alchemical Philosophers being an Attempt towards the Recovery of the Ancient Experiment of Nature* (London: T. Saunders, 1850), 143.

[2] Principe, *Secrets of Alchemy*, 98. [3] Principe, *Secrets of Alchemy*, 99.

[4] Principe, *Secrets of Alchemy*, 63–69.

[5] For valuable pioneering studies, see Lyndy Abraham, "Milton's 'Paradise Lost' and 'the sounding alchymie,'" *Renaissance Studies* 12, no. 2 (1998): 261–76, and Harinder Singh Marjara, *Contemplation of Created Things: Science in "Paradise Lost"* (Toronto: University of Toronto Press, 1992), 163–86.

as examples of Milton drawing from a ready-to-hand source for vitalist metaphors or as evidence of an alchemical cast to his natural philosophy? I will argue that Milton's use of alchemical ideas is precise and informed by a real engagement with alchemical principles, and furthermore that he had close connections to a network of friends engaged in alchemical pursuits and familiar with alchemical texts. Already in his early 20s Milton in an academic exercise speaks of "the pleasure…to be found in Study [of]…the hidden virtues of plants and metals."[6] Milton's declared interest in practical knowledge in *Of Education* makes it likely that he would be curious about those alchemical pursuits and texts, for alchemists in Milton's day were focused on practical applications in metallurgy and medicine.[7]

In this chapter I will briefly outline the alchemical tradition that Milton and Newton inherited, before addressing the vitalist turn in alchemy in the early modern period. Then, violating chronology, I will begin with Newton rather than Milton, for Newton's engagement with alchemy is more extensive and explicit than Milton's, and it provides a context that can illuminate Milton's own references to alchemy. Newton's alchemical writing and experimentation extended over decades, during which time his thinking evolved. I will not attempt to reproduce all the byways of Newton's understanding of alchemical operations; instead, I will focus on those central aspects that can shed light on Milton's own engagement with the art. Newton's serious engagement with alchemy can help us to discern when and how imagery in *Paradise Lost* that might seem fanciful or based on outmoded natural philosophy is instead philosophically current.

I will first address what can appear, and what used to appear to me, the dismissal of alchemy from Milton's own pen. When narrating Raphael's lunch with Adam and Eve in the Garden, Milton emphasizes the point that the angel actually consumes and metabolizes the food offered by Eve, thus distancing himself from prevailing dualist accounts that suggested that angels, lacking bodies, only appeared, by means of temporary bodies gathered from the air, to eat. He addresses readers' skepticism that an angel can eat and digest:

> nor wonder; if by fire
> Of sooty coal the empiric alchemist
> Can turn, or holds it possible to turn
> Metals of drossiest ore to perfect gold
> As from the mine. (5.439–43)

The narrator's "or holds it possible" marks his (and Milton's) skepticism. But skepticism of precisely what? Is the narrator skeptical of the potential for alchemical

[6] Milton, Prolusion VII, "Learning brings more Blessings to Men than Ignorance," *CPW* 1:295–96.
[7] See Milton's *Of Education*, *CPW* 2:389–96.

transformation of substances or skeptical of alchemists' success in discovering the philosophers' stone? On the sun, Milton's Satan finds stones resembling Earth's most brilliant gems, as well as

> a stone besides
> Imagined rather oft than elsewhere seen,
> That stone, or like to that which here below
> Philosophers in vain so long have sought,
> In vain, though by their powerful art they bind
> Volatile Hermes, and call up unbound
> In various shapes old Proteus from the sea. (3.598–604)

Milton distinguishes solar alchemy from the Earth-bound pursuits of empiric alchemists. The production of the philosophers' stone, the alchemical substance that held the key to the transmutation of substances (and thus for empiric alchemist the key to turning lead into gold), is figured as perennially elusive for terrestrial alchemists, but as occurring naturally on the sun. As Abraham suggests, Milton is not "questioning the validity of alchemy, only the possibility of the success of earthly alchemy."[8] The key to the transmutation of substances exists, even if not yet discovered here. Newton himself illustrates the distance between confidence in the possibility of producing the philosophers' stone and acknowledgment of those efforts as to date having fallen short. In "Humores minerales," an early alchemical manuscript on the verso of the folios of "Of natures obvious laws and processes in vegetation," discussed in the previous chapter, Newton notes that "vulgar chymists" (*vulgares chemici*) are unable to reproduce the natural process by which metals are vaporized and driven by the heat of the earth upward to the atmosphere before returning to earth to combine with sulfur to generate pure metals, but he does not rule out the possibility that alchemical adepts can achieve this result: "Metals may by no artifice be drawn into so great a volatility that they can ascend in the manner of vapor by means of a gentle terrestrial heat, unless they lose their metallic nature,... because the vulgar chymists cannot thus attenuate them—^this is the labor, this is the work."[9] The interpolated passage from the *Aeneid* 6.129, "this is the labor, this is the work/*hic opus hoc labor est*,"[10] marks

[8] Abraham, "Milton's 'Paradise Lost' and 'the sounding alchymie,'" 270.

[9] Newton, "Humores minerales," Dibner Ms. 1031, 6v; https://webapp1.dlib.indiana.edu/newton/mss/dipl/ALCH00081, accessed April 1, 2019. I quote the translation available with the text cited. The Latin text reads, "*Metalla nullo artificio* <illeg.> *in tantam volatilitatem* <illeg.> *reduci possint* <illeg.> *ut* <illeg.> *cum calore leni terrestri ad modum vaporis ascendant.* <illeg.> *nisi naturam metallicam amiserint. Tum quia Vulgares Chemici non possunt ita attenuare* ^*hic labor hoc opus est*." Further references to "*Humores Minerales*" are taken from this source and cited by folio number. For a helpful explication of this passage, see William R. Newman, *Newton the Alchemist: Science, Enigma, and the Quest for Nature's 'Secret Fire'* (Princeton: Princeton University Press, 2019), 141–42.

[10] Newton inadvertently transposes the pronouns in Virgil's "*hoc opus, hic labor est*."

Newton's acknowledgment of the distance between what occurs in nature and is theoretically possible in the laboratory, on the one hand, and what he has in fact achieved in his own laboratory, on the other. The Sibyl's warning to Aeneas signals the difficulty, not the impossibility, of Aeneas' return from his prospective visit to the underworld. Newton was guardedly optimistic that alchemical adepts, whom he hoped to join, could generate pure metals. In *Paradise Lost*, Milton imagines the philosophers' stone as existing in fact on the sun, even while glancing at the failure of so many on Earth to reproduce it.[11]

Early Modern Alchemy and Its Roots

Zosimos of Panopolis, who flourished around 300 CE, proposed that metals were produced from the interaction of sulfur and mercury. He followed the Book of Enoch in holding that the art of metallurgy was taught by fallen angels to their human wives, the "daughters of men," a mating described in *Paradise Lost* (11.573–92). Zosimos maintained that metals are composed of a body (*soma/σῶμᾰ*) of mercury and a spirit (*pneuma/πνεῦμα*) of sulfur. The mercurial body is common to all metal, from the most noble to the most common. Metals are distinguished by the particular nature of the sulfurous spirit and by the purity or impurity of the two components. This version of alchemy with its variations was influential up through and including the early modern period. Paracelsus in the sixteenth century added *sal* or salt as a third element composing metals.[12]

This mercury-sulfur model of the generation and composition of metals was compatible with a mechanist understanding of material change, which could come about through the addition or subtraction of pure or impure alchemical elements and other, foreign substances. But the model proved adaptable to the rise of vitalist alchemy in the early modern period. Alchemical thinkers increasingly posited the activity of vital spirits working not only in animals and plants but also in the elementary components of metals. As Newton laid out his understanding early in the 1670s in "Of natures obvious laws": "metalls vegetate after the same laws" as plants, according to "circumstances in which they agree with plants & animalls," and "vegetation is the sole effect of a latent spirit & that this spirit is the

[11] Lyndy Abraham (271) rightly notes that in *Milton among the Philosophers* I did not fully recognize early modern alchemy as an important context for Milton's animist materialism; Abraham also notes the importance of alchemy in Isaac Newton's thought. In *Marvell and Alchemy* (Aldershot: Scolar Press, 1990), Abraham provides a rich account of the prominence of alchemical thought in seventeenth-century England, involving figures close to Milton, including his intimate friend Lady Ranelagh, his physician Nathan Paget, and Samuel Hartlib, who solicited Milton's composition of *Of Education* (12–14). Writing before Principe (see note 1 above), Abraham subscribes to the mistaken view that alchemy was a mystical pursuit (16–26).

[12] I use the term "element" here loosely, and not in the modern sense of the elements of the periodic table, the list of substances that cannot be reduced into simpler substances by chemical reactions, which was a product of the later eighteenth century.

same in all things."[13] This latent spirit in metals plays the role of seminal fluid in animals. According to William Newman, a leading authority on Newton and early modern alchemy, "Newton is probably thinking of sexual generation across the three kingdoms [animal, vegetable, and mineral] of nature."[14] As minerals were produced by vital processes directly analogous to sexual generation in fauna and flora, so they were subject to decay and death, the full life cycle of living organisms. This vitalist or hylozoic understanding of minerals, Newman demonstrates, is not a holdover from earlier, pre-scientific animist beliefs; again, most medieval thinking about alchemy was rooted in a mechanist rather than a vitalist set of assumptions. Tracing the rise of vitalist thinking in alchemy to late sixteenth-century European miners and metallurgists, Newman argues that "Newton ... was the heir of a unique blend of mining lore and alchemy that had reached its efflorescence almost a century before."[15]

Robert Boyle indicated in 1666 the latitude of the term "life": "'tis ascribed not only to all sorts of Animals and Plants, but by many Chymists and Mineralists to Stones and Metals growing in the bowels of the Earth."[16] In portraying in *Paradise Lost* the animation of what was and still is usually thought of as inanimate, Milton reflected the thought of leading natural philosophers, including vitalist alchemical theorists.

Newton's thinking on alchemy, and especially on the alchemical vision of a living cosmos and Earth, is marked by the influence of a Polish alchemist, natural philosopher, medical doctor, and mining official who died six years before his birth, Michael Sendivogius, or Michał Sędziwój (1566–1636), and whose *Novum lumen chemicum* (1604) Newton acquired in the late 1660s.[17] Building on but diverging from medieval and more mechanist explanations of mineral as derived from the intermixing of sulfur and mercury, Sendivogius ascribed a special formative power to niter (potassium nitrate, also known as saltpeter and *sal nitrum*). He theorized that niter beneath the surface of the Earth attracts a vital "aerial niter" from the heavens. Niter's usefulness as a plant fertilizer and preserver of meats from decay was taken as a sign that it possessed a vital power.

Sendivogius' aerial niter worked in concert with the Paracelsan alchemical triad of mercury, sulfur, and salt. His sulfur and mercury are closely related; the alchemical sulfur is a more active form of the "philosophical mercury." The mercurial elements inject their seeds or *sperma* into the subterranean hollow or "womb" of the earth, where they are digested and released from their superfluous

[13] Newton, "Of natures obvious laws," 1r.

[14] Newman, *Newton the Alchemist*, 154.

[15] Newman, *Newton the Alchemist*, 13.

[16] Robert Boyle, *The Origins of Forms and Qualities*, in *The Works of Robert Boyle*, eds. Michael Hunter and Edward B. Davis, 14 vols (London: Pickering and Chatto, 1999), 5:467–68, quoted in Helen Thompson, *Fictional Matter: Empiricism, Corpuscles, and the Novel* (Philadelphia: University of Pennsylvania Press, 2017), 22.

[17] Newman, *Newton the Alchemist*, 136; my summary of Sendivogius' theory is drawn from this book, 67–69 and *passim*.

matter, which is cast off as stone. The seeds, vaporized and driven up by the heat of the interior central sun, rise toward the surface of the earth, where they combine with the philosophical sulfur in the soil and produce metals and minerals. The fewer the impurities, the nobler the metals. Those vapors not drawn into combination with the sulfur rise further to the surface of the planet, where condensing and combining with water they nourish plants. At the surface, the elemental sperm in the mercurial water is empowered by the celestial aerial niter and driven into the atmosphere, where it is further enlivened by the aerial niter, before returning, carried by rain and the sun's rays, to the Earth to join with sulfur to produce more niter or saltpeter. The career of the niter, sulfur, and mercury is thus cyclical, descending to and into the Earth before rising from within the Earth to the heavens, where the cycle begins again. This alchemical model is the natural philosophical foundation of the cosmical organic cycles that one finds in Milton's poetry and Newton's alchemical writings. Newman offers this condensed description of the process:

> The growth of metals in... mines is due to the same process as that of plants on the surface of the earth. Both depend on the descent of a vital power brought down by rain, which joins with the volatilized *sal nitrum*: the combination of this vital power and the *sal nitrum* acts like a sort of universal fertilizer. In this fashion, Sendivogius devised a cosmic system in which chymistry played the central role. The circulation of the aerial niter and its regeneration of the earth surely lie behind Newton's view of the tellurian globe as a living creature.[18]

Newton's Earth, as we saw in the preceding chapter, "resembles a great animall."

Sendivogius' picture of the vivifying circulation of the aerial niter finds an echo in the writing of the Paracelsan Robert Fludd (1574–1637). Fludd sought to isolate the vital spirit chemically, a sign that he did not view the *spiritus* or *pneuma* as immaterial. Like Sendivogius he identified this vital spirit with the aerial niter, and he believed himself to have distilled it from the germ of wheat.[19] For Fludd as for Sendivogius, the niter circulates, arising from within the earth to the surface and thence to the heavens, from whence it returns to the earth, enlivened by aerial niter, carried by the beams of the sun and by rain, where it is responsible for the generation and preservation of life. Fludd in his *Philosophicall Key* emphasized the parallels between the human body and the universe, related as microcosm and macrocosm. The first to defend in writing William Harvey's demonstration of circulation of the blood,[20] Fludd saw in it a confirmation of his

[18] Newman, *Newton the Alchemist*, 69.

[19] Allen Debus, "Chemistry and the Quest for a Material Spirit of Life in the Seventeenth Century," in *Spiritus: IVth Colloquio Internazionale*, eds M. Fattori and M. Bianchi (Rome: Edizioni dell'Ateneo, 1984), 245–63.

[20] Debus, "Chemistry and the Quest for a Material Spirit," 259.

argument for the circulation of the aerial niter, which descends from the sun to the earth and returns to be revivified from the earth to the sun, just as the arterial blood flows from the heart and returns to it as venous blood to be revivified. The purified philosophical niter, which for these alchemists was the vivifying element in the generation of metals, was active in all three kingdoms: animal, vegetable, and mineral. According to Allen Debus, Fludd maintained that "All three kingdoms of nature are dependent on it for life and we see its source in the sun"; in Fludd's own words, "all things [are] equally adorned and vivified by the influence of her beams."[21]

The search for the purified aerial niter amounted to the search for the philosophers' stone, that substance that by its purity and vital power could under the right circumstances transform base metals into gold. And given that it operated on all three kingdoms, including the animal (and thus human), it was also the search for the elixir, the universal medicine, that would reverse disease and restore health. Chrysopoeia and iatrochemistry, or medical alchemy, which we would separate under the non-contiguous domains of metallurgy and medicine, were aspects of the same study.

Newton and Alchemy

The influence of vitalist alchemy on Newton is apparent in two manuscripts introduced above, written on reversed sides of the same manuscript folios (Dibner 1031B), dating from the last years of Milton's life, the English "Of natures obvious laws and processes in vegetation" and the Latin "Humores minerales" or "Mineral Humors." In the Latin manuscript Newton addresses a problem raised by the evidence that metals are dissolved by descending corrosive waters, which miners observe as dripping moisture.[22] These descending waters, alchemists theorized, are vaporized by contact with the earth's interior heat; as a result, they ascend, leaving the metal behind, as demonstrated by the observable fact that the distillation of metalline solutions sublimes the solvent and leaves the metal behind. If the metal is left behind, Newton wonders, why is it the case that over time all the metal in the Earth has not been leached down and collected at the center, leaving the layers closer to the surface without metal: "Hence with the metals continually drawn downwards, never ascending so long as they remain

[21] Debus, "Chemistry and the Quest for a Material Spirit," 251, quoting from *Robert Fludd and His Philosophicall Key: Being a Transcription of the Manuscript at Trinity College, Cambridge*, with an Introduction by Allen G. Debus (New York: Science History Publications, 1979), 91.

[22] Newton, "Humores minerales," "humidities of this sort are found everywhere falling toward the center of the earth. For thus miners observe drops perpetually distilling through the walls of mines and they find the cavities of crags to be full of waters/Sic enim fossores observant guttas perpetim destillantes per latera fodinarū & cavitates rupium aquarū plenas <*illeg.*> inveniunt" Dibner MS 1031B, fol. 6^{v}, accessed July 10, 2025.

metals it would be necessary that in a few years the greatest part would have vanished from the upper earth." This obvious counterfactual leads in the next breath to an ultimately vitalist answer: "unless they are conceded to be generated there de novo."[23] His answer is that within the Earth, in a process supposedly theoretically replicable by adepts though not by "vulgar chymists" (nor yet by the aspiring adept Newton himself), metals are not merely dissolved but destroyed, leaving the released metallic sulfurous and mercurial spirits (*spiritus metallici*) to arise as fumes, at which point they meet descending metals dissolved mechanically, and induce the dormant spirits of the descending dissolved metals into a "state of motion and vegetation, . . . putrify[ing] and destroy[ing] the metallic form and convert[ing] it into spirits similar to themselves. Which can then ascend again and thus a perpetual circulation of metals takes place."[24]

These metallic spirits are then the seeds that not only generate, *de novo*, metals in the upper earth but also, given the continuity of mineral, vegetable, and animal kingdoms, "bestow life on animals and vegetables."[25] Newman relates the manuscript's description of the circulation of the most refined metallic spirits, the aerial niter, and its generative power to Newton's understanding of an ancient alchemical text, the *Emerald Tablet*, which Newton translated and upon which he commented.[26] The *Emerald Tablet* or Tabula Smaragdina is a brief, gnomic text attributed to Hermes Trismegistus; it was viewed in medieval and early modern Europe as a foundational alchemical text. Newton translates the opening lines thus:

> That which is below is like that which is above & that which is above is like that which is below to do the miracles of one only thing
>
> And as all things have been & arose from one by the mediation of one: so all things have their birth from this one thing by adaptation.
>
> The Sun is its father, the moon its mother, the wind hath carried it in its belly, the earth is its nourse. The father of all perfection in the whole world is here. Its force or power is entire if it be converted into earth.
>
> Separate thou the earth from the fire, the subtile from the gross sweetly with great indoustry. It ascends from the earth to the heaven & again it descends to the earth & receives the force of things superior & inferior.[27]

[23] "Hinc metallis continuò deorsu delatis et nunquam dum sunt metalla ascendentibus: necesse esset ut intra paucos annos maxima pars a superiori terrâ evanesceret. nisi ibi e novo generari concedantur," fol. 6^{v}.

[24] Newton, "Humores minerales," "in statu motus ut vegetationis, putrifaci ent, formam metallicam destruit <*sic*> et <*illeg.*> in spiritus sibi similes vertunt. Qui rursus tunc possunt similiter ascendere et sic fit continua circulatio metalloru," fol. 6^{v}.

[25] Newton, "Humores minerales," "animalibus et vegetabilibus vitam largitentur," fol. 6^{r}.

[26] Newman, *Newton the Alchemist*, 145–47.

[27] Newton, Keynes Ms. 28, 2^{r}; https://webapp1.dlib.indiana.edu/newton/mss/norm/ALCH00017, accessed January 22, 2024.

Here is a picture of one first matter circulating from heavens to the Earth and back to the heavens, in a dynamic resembling what one finds in *Paradise Lost*.

"Of natures obvious laws" elaborates on a distinctive element of Newton's alchemical thought also present in the "Humores minerales," the complementary operations of vitalist and mechanist elements in alchemical transformation. Only after bodies were separated by mechanical means into sufficiently fine and vaporous parts could the released spirits vegetate and generate new metals and enliven plants and animals. Subtle spirits are latent or dormant in bodies until released first by mechanical separation and then animated by alchemical vital fermentation, at which point they can foster new life in the form of minerals, vegetables, and animals. Below I will suggest that Newton's distinction between active and latent or dormant spirits can help us to understand the otherwise puzzling dichotomy between the living and the non-living attributed by Raphael to the "one first matter," the "various degrees / Of substance, and in things that live, of life" (*PL* 5.473–74).

Newton begins "Of natures obvious laws" with observations about metal dendrites, tree shaped formations that formed in glass flasks; these dendrites suggested to him as to other early modern alchemists that metals vegetated: "sometimes metalls grow like trees in the earth."[28] He continues on the same page that "That vegetation is the sole effect of a latent spirit & that this spirit is the same in all things only discriminated by its degrees of maturity." As in "Humores minerales," this vegetation embraces the three kingdoms of animal, vegetable, and mineral, and the spirit organizes and enlivens matter, the process beginning with the "putrefaction or fermentation whereby there is an intimate union & exertion of Spirits & purgation of impuritys." As there are cycles of descent and re-ascent of the aerial niter, so there are cycles of dissolution of bodies into "black stinking rottennes" and regeneration into organized forms. All alchemical transformation first involves resolution into formlessness, which can then be re-formed by the vital spirit or aerial niter. Significantly, another name for the formless rottenness is *chaos*; Newton writes that nature's

> first action is to blend & confound mixtures into a putrifyed Chaos
> Then are they fitted for new generation or nourishment
> All things are corruptible
> All things are generable[29]

I will return below to similarities between Milton's chaos ("The womb of Nature and perhaps her grave" [*PL* 2.911]) and Newton's choice of a name for putrefied or disaggregated matter.

[28] The quotations in this paragraph, except for the last, can all be found in Newton's "Of natures obvious laws," 1^r.

[29] Newton, "Of natures obvious laws," 5^r.

Newton goes on to elaborate on the action of the air and aether in the circulation of the vital spirit outlined in "Humores minerales." The rising of air and vapors leaves a gap filled by the more tenuous, and thus faster, aether, which, descending, drives the vital spirits downward:

> Soe much Aether ought to descend as air & exhalations ascend, & therefore the Aether being by many degres more thin & rare then air (as air is then wather [sic]) it must descend soe much the swifter & consequently have soe much more efficacy to drive bodys downward then air hath to drive them up. And this is very agreeable to natures proceedings to make a circulation of all things. Thus this Earth resembles a great animall or rather inanimate vegetable, draws in aethereall breath for its dayly refreshment & vitall ferment & transpires again with gross exhalations.[30]

When claiming that the "Earth resembles a great animall or rather inanimate vegetable," how does Newton negotiate the relation of animate and inanimate? He seems here to suggest that the Earth may not be an animal with its senses and its reason, whether higher human reason or lesser animal reason, but that it is closely akin to a vegetable and similarly alive, that it, in other words, has the equivalent of a (material) vegetable soul, if not a (material) animal soul.

In describing the "subtil spirit which searches the most hiden recesses of all grosser matter" as "Natures universall agent, her secret fire, the onely ferment & principle of all vegetation. The material soule of all matter,"[31] Newton suggests that the material living principle is not divided dualistically from matter like the soul of dualist, but is itself subtly material. Newton's description here recalls Milton's one first matter, which can be "more refined, more spiritous, and pure" (*PL* 5.475). In its picture of the material soul's descent, mixing with less tenuous matter, and re-ascent after dissolution, it recalls even more the destination of the material soul in the *De Doctrina Christiana*, where, as we saw in the previous chapter, Milton speculates, citing Euripides' *Suppliants*, that the corporeal spirits return at death to the "four winds" (*DDC* I.13; *OCW* 8.1:455). In both Newton and Milton, the material soul, not only of minerals and plants but also of the human being, is not immortal; instead it is dispersed at death and dissolution, only to be reconstituted at the Last Judgment. In both again, the more tenuous substance ascends into the air and the heavens. Newton's material soul accounts for all action (as opposed to the motions arising passively as the result of mechanical collision) of those living, extra- or sub-mechanical events. The vital spirits make up this soul, and the "principles of her vegetable actions are noe other then the seeds or seminall vessels of things those are her onely agents, her fire, her soule, her life."[32]

30 Newton, "Of natures obvious laws," 3^{v}.
31 Newton, "Of natures obvious laws," 3^{v}.
32 Newton, "Of natures obvious laws," 5^{r}.

Newton elaborates on the alchemical circulatory system of the Earth in a text addressed briefly in the preceding chapter, the "Hypothesis of Light," a letter sent in December 1675, thirteen months after Milton's death, to the Royal Society by way its secretary, Milton's friend Henry Oldenburg. Newton here addresses the existence, motion, and effects of æthereal spirits, which are finer than æreal spirits, as the æther is of the "same constitution with air, but far rarer, subtler, and more strongly elastic."[33] These æthereal spirits are the source of vital motions in the æther. They rise like the æreal niter of "Of natures obvious laws" into the atmosphere, where some condense and descend, a motion that Newton at this time speculates may be the cause of gravity.[34] These spirits also ascend to the sun and the planets, and the circulatory system of the Earth is one part in the larger cosmic circulatory system:

> for nature is a perpetual worker, generating fluids out of solids, and solids out of fluids, fixed things out of volatile, and volatile out of fixed, subtil out of gross and gross out of subtil; some things to ascend, and make the upper terrestrial juices, rivers, and the atmosphere; and by consequence, others to descend for a requital to the former. And, as the earth, *so perhaps may the sun imbibe this spirit copiously*, to conserve his shining, and keep the planets from receding further from him. And they, that will, may also suppose, that *this spirit affords or carries with it thither the solary fewel* and material principle of light: and that the vast æthereal spaces between us and the stars are for a sufficient repository for this food of the sun and planets. (251; my emphasis)

Here again we see Newton echoing Milton's image of the sun "sup[ping] with the ocean," that the food of the living sun rises from a living earth.

Paradise Lost, Newton, and Alchemy

What are we to make of the uncanny similarity between Milton's and Newton's descriptions of the cosmic cycle of animate feeding? Milton's earth and moon "nourishment exhale /... to higher orbs,"[35] while what Newton's Earth consumes daily in "æthereall breath" it "transpires again with gross exhalations." Milton's

[33] Newton, "Hypothesis Explaining the Properties of Light, 250. http://www.newtonproject.ox.ac.uk/view/texts/diplomatic/NATP00002, accessed April 12, 2019. Further references to this text will be taken from this source and indicated by page number in Thomas Birch, *The History of the Royal Society*, vol. 3 (London, 1757).

[34] While there is continuity in the idea of a circulatory system, Newman argues that by this point Newton no longer accepted Sendivogian aereal niter as the vital principle, substituting for it sulfur (Newman, *Newton the Alchemist*, 442–44).

[35] The passage I quote has the moon exhaling, but the sense of the passage is that the moon's action is continuous and extends the Earth's, sending moist spirits upward to feed celestial objects.

sun "Sups with the ocean," while Newton's sun may "imbibe this spirit copiously." While Milton's probable and Newton's amply demonstrated interest in alchemical thinking may be sufficient for these echoes, it would be a mistake to discount the very real possibility of Newton's acquaintance with Milton's thinking or even his words, given the multiple links between their friends and acquaintances as well as close parallels in expression. In the light of Newton's statement, Milton's picture of the sun's supping with the ocean appears as a natural philosophical claim rather than as a metaphysical conceit.

Milton's network of friends and interlocutors, as I indicated at the outset, intersected with Newton's. Here I want to point to a network of acquaintance particularly relevant to alchemy. Newton most likely obtained at least some of the many alchemical texts he read and annotated from the circle of Milton's friend Samuel Hartlib.[36] Another key connecting figure is Nathan Paget, with whom Milton was acquainted by the early 1640s,[37] who was Milton's personal physician by the mid-1650s, and who by the early 1660s was Milton's neighbor on Jewin Street in London. It was Paget who introduced to Milton the young Quaker Thomas Ellwood, who credits himself with planting the seed in Milton's mind for *Paradise Regained*. Paget collaborated with the vitalist and medical alchemist Francis Glisson, Newton's Royal Society colleague.[38] Glisson, as we shall see, devised a theory of the blood's active self-sublimation into a form reminiscent of the sublimation of "intellectual spirits" described by Raphael in *Paradise Lost*. Given Milton's intense interest in medicine, occasioned by his blindness and digestive complaints, it would be surprising if Paget did not discuss medical alchemy with his curious patient.

Paradise Lost contains alchemical images and ideas, in marked contrast to the absence of alchemical references in the earlier poetry. This development may have something to do with genre; as expected of an epic poet, Milton aspires to display encyclopedic knowledge, and in *Paradise Lost* his ambitions are Lucretian as well as Homeric and Vergilian. At the same time, *Paradise Lost* is the first poem, aside from occasional and topical sonnets, to come after Milton's acquaintance with Hartlib, Paget, and Henry Oldenburg. The epic is rich in descriptions of an animate Earth and the organic generation of metals and gems, as well as passages employing technical alchemical ideas and terms. As noted above, the description of the philosophers' stone seen by Satan on the sun shares with Newton confidence in the existence of that object even while gesturing to failures to date to replicate it on Earth. If "here below / Philosophers" have not yet succeeded, the

[36] Richard Westfall, *Never at Rest: A Biography of Isaac Newton* (Cambridge: Cambridge University Press, 1980), 288.

[37] David Masson, *The Life of John Milton: Narrated in Connexion with the Political, Ecclesiastical and Literary History of His Time*, 7 vols (London, 1881–94; rpt. Gloucester, MA: Peter Smith, 1965), 6:455.

[38] William Riley Parker, *Milton: A Biography*, 2 vols, continuously paginated (Oxford: Oxford University Press, 1968), 979, n. 106.

theoretical aim is expressed in terms similar to Newton's and other alchemists'. They seek to "bind / Volatile Hermes," or fix volatile Mercury, the substance underlying alchemical transformation (*PL* 3.601–5).[39] What the alchemists hope to achieve by sublimation in his heated alembic (the "limbec" of *Paradise Lost* 3.605) occurs spontaneously, but with the same ingredients and by the same principles, on the sun. Milton also points in this passage to the alternate form of the philosophers' stone, the "elixir pure" or "Potable gold" (*PL* 3.607–8), which was believed to be a life-giving panacea. As in Newton, Sendivogius, and the other vitalist alchemists, the organic transformations unfold under the warming and vitalizing influence of the sun's rays.

Milton repeatedly points in the epic to alchemical conceptions of the vitalizing and impregnating power of the sun. The narrator makes the alchemical provenance explicit with a reference to the "arch-chemic sun" during Satan's Book 3 visit to the orb. Adam and Eve retire from their labors to their "cool bow'r" at mid-day, now "while now the mounted sun/Shot down direct his fervid rays to warm/ Earth's inmost womb" (5.300–302). When Adam asks his presumptuous question about the design of the universe in Book 8 ("reasoning I oft admire, / How nature wise and frugal could commit / Such disproportions" [8.25–27] as to have the vast array of stars serve the tiny and dark Earth), Raphael replies with a vitalist, alchemical lesson on the relation of sun and Earth:

> the Earth
> Though, in comparison of heav'n, so small,
> Nor glistering, may of solid good contain
> More plenty than the sun that barren shines
> Whose virtue on itself works no effect,
> But in the fruitful Earth; there first received
> His beams, unactive else, their vigor find. (8.91–97)

The sun, as in the alchemical literature, is masculine, one of the "two great sexes animat[ing] the world" (8.151), and cannot procreate without the female Earth. Adam makes the same point about the masculine procreative stellar power when Eve asks why the stars shine at night when they are asleep:

> these soft fires
> Not only enlighten, but with kindly heat
> Of various influence foment and warm,

[39] For Newton on volatile mercury see *Opticks: Or, A Treatise of the Reflections, Refractions, Inflexions and Colours of Light. Also Two treatises of the Species and Magnitude of Curvilinear Figures* (London: 1704), 61–63; https://www.newtonproject.ox.ac.uk/view/texts/diplomatic/NATP00037, accessed April 4, 2022.

Temper or nourish, or in part shed down
Their stellar virtue on all kinds that grow
On earth, made hereby apter to receive
Perfection from the sun's more potent ray. (4.667–73)

The lesser heat and less powerful rays of the more distant stars predispose the animal, vegetable, and mineral worlds to respond to the sun's "more potent ray." The idea that "all kinds" "receive / Perfection" from the sun's rays expresses the alchemical idea that the spirits descending with the sun's rays and moved by the sun's heat have the power to generate metals such as gold free from impurities.

The feminine procreative potency of the Earth features prominently in Milton's account of creation:

The Earth was formed, but in the womb as yet
Of waters, embryon immature involved,
Appeared not: over all the face of Earth
Main ocean flowed, not idle, but with warm
Prolific humor soft'ning all her globe,
Fermented the great mother to conceive,
Satiate with genial moisture. (7.276–82)

The Earth is both formed "in the womb" of waters and it comes itself to contain a womb, saturated with "genial moisture" that allows her to "conceive." The key verb here is "Fermented," an alchemical term for the vital transformation of substances. For Newton, nature's active laws account for fermentation, as opposed to the mechanical rearrangement of particles governed by nature's passive laws.

Fermentation is only one of the many technical alchemical terms in *Paradise Lost*. Vitalist Paracelsian alchemy lies behind the remarkable moment when Satan's headlong plunge through Chaos is arrested and reversed:

all unawares
Flutt'ring his pennons vain plumb down he drops
Ten thousand fathom deep, and to this hour
Down had been falling, had not by ill chance
The strong rebuff of some tumultuous cloud
Instinct with fire and niter hurried him
As many miles aloft. (2.932–38)

The "rebuff" here is the product of the æreal niter, which, as Sendivogius argued, is analogous to terrestrial *sal nitrum* or saltpeter, and it is both an essential component of gunpowder and, in the alchemical literature, a principle of life. Milton's contemporaries attributed lightning and thunder to the explosions sparked by the aerial niter.[40] Milton's alchemically literate first readers would have registered in the episode both the association with gunpowder of the arrest of Satan's perpetual plunge and the operation even in Chaos of the circulatory system of the created cosmos. That which arrests Satan's fall and thus takes part in the divine providential plan is part of the picture of a living matter common to Milton and Newton. The devils, as we will see, ascribe phenomena to an older, possibly mechanist alchemy.

The newer, vitalist alchemy may lie behind Milton's description of the Son's chariot of paternal deity, "Itself instinct with spirit" (6.752). If, as I have argued, Milton denies the existence of purely incorporeal substance, what are we to make of this spirit animating the chariot? This spirit may be related to, or an instance of, the spirit latent in all matter in the Sendivogian and Newtonian alchemical model. Latent spirits in earthly bodies are activated by spirits descending from the heavens. An empyreal object, the chariot exhibits this spirit as active and vividly perceptible rather than latent and inactive.

Satan in Book 6 observes, without stressing its vital nature, the activity of the alchemical exhalations of the living ground. "Gems and gold," he says, are the product of "materials dark and crude / Of spiritous and fiery spume," which, when "touched / With Heav'n's ray... shoot forth / So beauteous, op'ning to the ambient light" (6.475–81). The "spiritous and fiery spume" is, once, again, the alchemical æreal niter and sulfur. In his 1695 *Annotations on Milton's "Paradise Lost,"* Patrick Hume points to the vitalist alchemical freight of these lines:

> V. 478. *Materials dark and crude,* &c. *Deep underneath this Heavenly Soil there lie, hidden materials raw and undigested, of thin and fiery Foam, till toucht and warmed with Heavens bright Beams, they gain more perfect Being; and call'd forth by his cheering Ray, they spring enlivened, spreading into open day.... Spume;* Of the Lat. *Spuma,* froth, foam, a word expressing well the *crude consistence of Sulphur,* and other Subterranean Materials, the efficients of Fertility. This is a Description of Salt Petre.[41]

Hume employs alchemical terms—sulfur, saltpeter—in a description of an alchemical process of fertilization. What Milton has Satan witness are the "exhalations" of volatile elements, which early modern miners searched for as signs of underground deposits of ore. The colors of these flashing exhalations were taken by the author of the sixteenth-century alchemical text *De natura rerum* to reveal the nature of the buried ore, white for tin, lead, or silver; red for copper; and yellow

[40] Allen G. Debus, "The Paracelsian Aerial Niter," *Isis* 55, no. 1 (1964): 43.
[41] Patrick Hume, *Annotations on Milton's "Paradise Lost"* (London, 1695), 202; Early English Books Online, ProQuest doc. 2240982831, accessed April 16, 2019.

for gold.[42] Newton mentions this idea in the manuscript "Notes on the mining, preparation and properties of 'Saturn'": "Corruscation like Gunpouder running along is a signe of metals unripe & in primo ente."[43] According to pseudo-Paracelsus, these coruscating exhalations are signs not of perfected metals but of the gestation of metals: "as long as those Coruscations appear, whether they be great, or small, or of this or that colour, the Metalls in those Mines have not yet attained to perfection, but are yet in their first being as the sperm of a man in the matrix of a woman."[44] The text refers here to the sexual generation of metals, an understanding shared by Newton and Milton and early modern vitalist alchemists more generally. Milton has Satan gesture toward this understanding when suggests that the ores "in their dark nativity.../ Shall yield us pregnant with infernal flame" the materials of war (6.482–83).

Milton describes another and related sign of what was taken to be the living generation of metals when describing the discovery in Hell of the materials for the construction of Pandemonium:

> There stood a hill not far whose grisly top
> Belched fire and rolling smoke; the rest entire
> Shone with a *glossy scurf*, undoubted sign
> That in his womb was hid metallic ore,
> The work of sulfur. (1.670–74; my emphasis)

The *OED* cites this passage as an instance of the definition of "scurf" (n. 3): "Any incrustation upon the surface of a body;...a saline or sulphurous deposit, mould, or the like." Shining blue or green deposits of copper and iron sulfates coating the walls of mines were viewed by Sendivogius and others as evidence of the vital generation of metals, a conclusion supported by the fact that these deposits were seen to grow and change over time.[45]

Despite observing a phenomenon that vitalist alchemists viewed as living, Satan immediately puts a mechanist spin on the dark materials, suggesting that he has been speaking metaphorically of a "pregnant" earth a line before:

> Which into hollow engines long and round
> Thick-rammed, at th' other bore with touch of fire

[42] On this point, Newman (*Newton the Alchemist*, 79) cites Pseudo-Paracelsus, *De natura rerum* (see n. 44 below).

[43] Quoted in Newman, *Newton the Alchemist*, 79.

[44] *Of the Nature of Things, Nine Books*, Written by *Philipp Theophrastus* of *Hohenheim*, CALLED *PARACELSUS*, tr. J.F., M.D. (London, Printed by *Richard Cotes*, for *Thomas Williams*, at the Bible in Little-Britain, 1650), 130; Early English Books Online, ProQuest doc. 2264207732, accessed August 7, 2025.

[45] Newman, *Newton the Alchemist*, 71. See Color Plate 2 for a photograph of what can be described as a bright blue "glossy scurf" of copper sulfate coating the walls of a mine in Cornwall.

Dilated and infuriate shall send forth
From far with thund'ring noise among our foes
Such implements of mischief as shall dash
To pieces, and o'erwhelm whatever stands
Adverse, that they shall fear we have disarmed
The thunderer of his only dreaded bolt. (6.484–91)

As I have argued in the past, the thunderer's *bolt*, the living weapons of God, like the living chariot, are simultaneously and inseparably material and spiritual, "infix[ing] / Plagues" in the rebel angels and leaving them with inextricably corporeal and spiritual wounds, "Exhausted, spiritless, afflicted, fallen" (6.837–38, 852).[46] Satan, on the other hand, devises a weapon designed to put matter in motion, to "dash / To pieces" larger bodies into smaller bodies, precisely the kind of macroscopic effect of Newton's passive, mechanical laws, which operate above the subtler and less visible active laws that govern the vegetation of material substances. The hollowness of Satan's mechanical weapons might suggest the absence of material active spirits.

The "glossy scurf" discussed above is an "undoubted sign" of the "work of sulfur" in the "womb" of the material heaven. This is the alchemical sulfur that activates the latent mercurial spirits in immature matter, or the ores of perfected metals. Satan, however, chooses to work only mechanically with potentially living elements. To fashion his hellish artillery in Heaven, the devils

turned
Wide the celestial soil, and saw beneath
Th' originals of nature in their crude
Conception; sulfurous and nitrous foam
They found, they mingled, and with subtle art,
Concocted and adusted they reduced
To blackest grain, and into store conveyed:
Part hidden veins digged up (nor hath this Earth
Entrails unlike) of mineral and stone,
Whereof to found their engines and their balls
Of missive ruin. (6.509–19)

The "sulfurous and nitrous foam" are the æreal niter that Newton first took as carrying the vital activating spirits and the sophic sulfur that came to take niter's place in Newton's alchemical thinking. But there is no hint in the devils' art of the sublimation, rarefaction, and perfectibility of matter nor the quickening into life

[46] Stephen M. Fallon, *Milton among the Philosophers: Poetry and Materialism in Seventeenth-Century England* (Ithaca: Cornell University Press, 1991), 238–39.

that is the hallmark of Sendivogian or Newtonian vitalist alchemy. Instead, the devils reduce the material to a gross "blackest grain" to be warehoused until destructive use. Ironically, given Satan's boast of being "self-begot, self-raised/By our own quick'ning power" (5.860–61), Satan is blind to the latent life secreted as *semina* in all matter. The sulfur of pre-Sendivogian alchemy was viewed as instrumental in making pure metals by a combination with pure mercury in the proper proportions, not as a living principle activating in sexual or quasi-sexual terms latent life in the mercurial receptacle. The infernal recipe is evocative of earlier, mechanist alchemy, and Satan's occluding of vitalist elements fits his sterile view of God's creation with its material life.

The distance between the vitalist alchemical view of the world that Milton shared with Newton and its Satanic alternative can be measured by contrasting Milton's descriptions of the creation of the world on the one hand and the building of the bridge over Chaos on the other. The keynotes of Milton's narration Sin and Death's construction of the bridge are lifelessness and the forceful moving of parcels of matter:

> Hovering upon the waters; what they met
> Solid or slimy, as in raging sea
> Tossed up and down, together crowded drove
> From each side shoaling towards the mouth of Hell.
> As when *two polar winds* blowing adverse
> Upon the Cronian Sea, together drive
> Mountains of Ice....
> ....
> The aggregated soil
> Death with his *mace petrific*, *cold* and *dry*,
> As with a trident smote, and fixed as firm
> As Delos floating once; the rest his look
> Bound with Gorgonian rigor *not to move*,
> And with asphaltic slime; broad as the gate,
> Deep to the roots of Hell the gathered beach
> They fastened, and the mole immense wrought on
> Over the foaming deep high-arched, a bridge
> Of length prodigious joining to the wall
> Immovable of this now fenceless world
> Forfeit to Death. (10.285–304; my emphasis)

Instead of an alchemical subliming of pure metals from stone, or a Sendivogian process in which undigested traces of impure are cast off as stone, Death with a "mace petrific" turns unformed matter to stone. Instead of dynamic, metabolic movement, we have absolutely static matter, "Bound with Gorgonian rigor not to move." In place of the liquid activating warmth of alchemy, the mace is "cold and dry," and Milton gives us "two polar winds" driving "Mountains of ice." Instead of a living world, we have a world "Forfeit to death."

Signaling the contrast between the Son's creation of a living universe in Book 7 and the moribund anti-creation in Book 10, Milton depicts Sin and Death "Hovering upon the waters," in a parody of "the Spirit of God mov[ing] upon the face of the waters" (Genesis 1:2). The action of the divine creative spirit is organic and vital rather than mechanical. As such it resembles the creation as interpreted by early modern alchemists. Lyndy Abraham observes that in *The History of the World* (1614) Sir Walter Raleigh treats the creation story of Genesis alchemically, as God worked with the first matter, by, in Abraham's summary, "the condensation, rarefaction and conversion of the elements...and the separation and digestion of the earth from the waters."[47] Ralegh's picture complements the position of the *Emerald Tablet* as later translated by Newton. The first matter condensed, rarefied, converted, and separated corresponds to the "one thing" from which in Newton's translation "all things have their birth...by adaptation," and by "Separat[ion of] the earth from the fire, the subtile from the gross sweetly."[48] As outlined in Chapter 1, this first matter is a sensible continuing material substrate, not Aristotle's qualityless, insensible substance.

Newton, who pursued natural philosophy in the service of religion, as an instrument of celebrating divine omnipotence, added his voice to those viewing creation as an alchemical process. According to his Commentary on the *Emerald Tablet*,

> just as the world was created from dark Chaos through the bringing forth of the light and through the separation of the aery firmament and of the waters from the earth, so our work brings forth the beginning out of black Chaos and its first matter through the separation of the elements and the illumination of matter. Whence arise the marvelous adaptations and arrangements in our work, the mode of which here was adumbrated in the creation of the world.[49]

The "black Chaos" of "our work" (or alchemy) is the black stinking rottenness that results from the putrefaction of vegetated matter (i.e., all matter) as outlined in "Of natures obvious laws." The dark Chaos from which God creates corresponds

[47] Lyndy Abraham, "Milton's 'Paradise Lost' and 'the sounding alchymie,'" 262, citing Walter Raleigh, *The History of the World: In Five Books* (London, 1614).

[48] Newton, Keynes Ms. 28.

[49] Newton, "Hermes," 6^v, Keynes Ms. 28, translated by Betty Jo Teeter Dobbs in *The Janus Faces of Genius: The Role of Alchemy in Newton's Thought* (Cambridge: Cambridge University Press, 1991), 277, 277 (and quoted on 68). For the transcription of the Latin original, see Dobbs (276) or the text on the *Chymistry of Isaac Newton*, as here: "*Et quemadmodum mundus ex Chao tenebroso per lucis productionem separationem firmamenti aerie et aquarum a terra, creatus fuit sic opus nostrum ex Chao nigro & materia sua prima per separationem elementorum et illuminationem materiae originem ducit. Unde oriuntur adaptiones et dispositiones mirabiles in opere nostro quarum modus est hîc in mundi creatione adumbrates,*" http://webapp1.dlib.indiana.edu/newton/mss/norm/ALCH00017, accessed April 17, 2019.

to this black Chaos. In linear Christian history, there is one creation from Chaos, and the world from Newton's perspective will age and begin to decay on its way to the final conflagration. In the meantime, elements in the world move through cycles of vegetation toward perfection and putrefaction back to black rottenness. Lives of individuals (persons, plants, minerals) are born, mature, decay, and die in succession. The world as a whole is created once, matures, and will decay and die.

As noted above, there is a strong alchemical cast to Milton's recounting of the first moments of creation:

> Thus God the heav'n created, thus the earth,
> Matter unformed and void: darkness profound
> Covered th' abyss: but on the wat'ry calm
> His brooding wings the Spirit of God outspread,
> And vital virtue infused, and vital warmth
> Throughout the fluid mass, but downward purged
> The black tartareous cold infernal dregs
> Adverse to life: then founded, then conglobed
> Like things to like, the rest to several place
> Disparted. (7.232–41)

Taken by itself, "Matter unformed and void" need not be construed alchemically, but we move into alchemical territory as God "infuse[s]" "vital virtue . . . and vital warmth / Throughout the fluid mass," and even more so as the spirit "downward purged / The black tartareous cold infernal dregs." Alchemical vegetation (and thus the growth and perfecting of substances) require both gentle heat and vital spirits, the vital virtue and vital warmth in this passage. Alchemical vegetation invariably involves separation and the purging of impurities.

Alchemy and Milton's "Infernal Dregs"

John Rogers has raised the question of whether the "infernal dregs / Adverse to life" represent an eruption of dualism into what is otherwise a vitalist monism in *Paradise Lost*, and a premature introduction of evil and death into the world before the Fall.[50] Paracelsus, Rogers notes, viewed tartareous dregs as the result of the Fall and the cause of disease. The problem, if it is one, is both natural philosophical and theological. If all matter is alive or potentially alive, where would be the place for dregs "adverse to life"? And if they were thought to appear at the Fall, what are they doing at creation?

[50] John Rogers, *Matter of Revolution: Science, Poetry, and Politics in the Age of Milton* (Ithaca: Cornell University Press, 1996), 134.

The answer to the theological question may be the simpler of the two. Charlotte Nicholls observes that, while the fall of Adam and Eve is yet to occur by the time of the purging of the dregs with the creation of the cosmos in Book 7, the rebel angels have already fallen, and Hell has been created.[51] Nicholls points to a key passage in the chapter on creation in *De Doctrina Christiana*: "it is neither matter nor form which sins, but yet, when either of them has issued from God and become another's property, what prevents it (being now mutable) from becoming infected and polluted through the calculations of the Devil and of mankind, which proceed from matter and form?" (*DDC* I.7; *OCW* 8.1:293–95). The devils are at home in a Hell that is a "universe of death, which God by curse / Created evil, for evil only good, / Where all life dies, death lives" (2.622–24). Upon their final defeat in the war in Heaven, they choose this place of death. While the narrator first indicates in the beginning of the epic that "Him [Satan] the Almighty Power / Hurled headlong flaming from th'ethereal Sky" (1.44–45), in Book 6 Raphael tells Adam and Eve that "headlong themselves they [the devils] threw / Down from the verge of Heav'n" (6.864–65). Still to be constructed, in part out of "asphaltic slime" (10.298), is the bridge over Chaos from Hell to Earth. One response to this sequence of events, and to Rogers' concern about the integrity of Milton's monism, is that the poem here shows the strain of turning natural philosophy into narrative. Having committed himself to a metaphysics of living matter, Milton admits an inconsistency into his poem in order to associate the devils with death. But another response is that the sin of the devils prompts the (ironic) generation of deathly matter. In Nicholls' fine articulation, the devils' "darkness is, in this natural philosophy, utterly their own, inexplicable, but signifying the potential of the creations of God to reject divinity."[52]

An answer to Rogers' concern is available also from within natural philosophy. Tartareous dregs are familiar elements in vitalist alchemy. An illuminating context for the dregs of Milton's creation scene can be found in the work of Heinrich Nolle, a medical alchemist who flourished in the second decade of the seventeenth century. Henry Vaughan, the poet and physician and brother of the alchemical writer Thomas Vaughan, translated Nolle's *Hermetical physick*; the translation was published in 1655 by Humphrey Moseley, who a decade earlier had published of Milton's *Poems… Both English and Latin*. Nolle distinguishes two varieties of tartareous dregs in wine, one vital and the other not. One, found in young wine, is composed particles that settle to the bottom over time, "as the grosser, earthy, and more impure part subside into a feculent substance, found alwaies in the bottome, and called Dregs," or the tartar separated over a longer

[51] Charlotte Nicholls, "Body Out of Spirit: Medical Science and the Creation of Living Soul in Paradise Lost," *Milton Studies* 57 (2016), 145.
[52] Nicholls, "Body Out of Spirit," 145.

time and found "stick[ing] to the Dregs or Lees of old Wine-pipes."[53] These dregs are cold, like the dregs in Milton's creation account. The other tartareous dregs are warm and vital:

> But I meane that *Tartar* also, which is in perpetual liquefaction and commixture with the most refined wines, and which gives them their tincture either red or any other. This true *Tartar,* either by Evaporation, or simple distillation...*is easily discerned to be moderately hot* [my emphasis], for the more liquid part of the humour (which was the *Vehiculum,* in which the *Tartar* in its dissolution was contained) being separated from it, the *Tartar* alone remaine in the bottom. This liquid humour, though of red wine, distills all bright and limpid, but the heavier red substance, which I call *Tartar,* stayes all behind: a solid substance....Nor is this *Tartar* onely in red, or white Wines, but in any other though decocted and also in the humours of mans body.[54]

This warm tartar that accounts for the vegetable red color of red wine is also present in our bodily humors. This tartar, nevertheless, distilled out of the blood, can lead to gout and stones, becoming as hard and stubborn as the cold dregs. Milton's "tartareous cold infernal dregs," thus, are familiar in vitalist alchemical medicine, and the grosser, more stubborn, and less than salutary elements have a place in a vitalist natural philosophical context.

The inconsistency that Rogers identifies is a function of Milton's invoking the processes of medical alchemy for instrumental purposes, with the dangerous by-products excluded from the creation of the unfallen universe and becoming available for infernal construction. While Rogers is correct that the dregs present an anomaly in a universe otherwise composed of living or potentially living matter, they serve a narrative purpose in the poem, as a possible material for the construction of Hell and, later, the bridge over chaos. To the extent that there is an inconsistency here, it points once again to a significant difference between the handling of substance in Milton and Newton, despite so many points of similarity. As a theoretical and experimental natural philosopher, Newton ties himself to strict canons of demonstration and consistency. As phenomena support or falsify his models, Newton maintains or changes them. Despite the impressive consistency and coherence of *Paradise Lost*, Milton is a poet looking for analogies to familiar experiences to communicate his vision of the cosmos of the poem. Looking for the material of Hell, Milton turns to tartareous dregs, to waste substance implicated in disease and death. Recall the description of Hell as a place that "God by curse / Created evil, for evil only good, / Where all life dies,

[53] Heinrich Nolle, *Hermetical Physick: or, The Right Way to Preserve, and to Restore Health*, trans. Henry Vaughan (London: Humphrey Mosely, 1655), 42–43; Early English Books Online, ProQuest doc. 2264207732, accessed July 10, 2025.

[54] Nolle, *Hermetical Physick*, 43–44.

death lives" (2.622–24). Milton's poetic vision of Hell is as an anomalous, paradoxical world where opposites enact each other: life dies, and death lives. This gives us other ways to think of the dregs adverse to life that preserve Milton's vitalism: they are the continuous dying of that which remains alive, or the immanent life even in death.

The separation of the "dregs / Adverse to life" from the material out of which the cosmos is born is only the first of the alchemical separations at creation. Immediately after the purging of the dregs, God "founded, then conglobed / Like things to like, the rest to several place / Disparted, and between spun out the air, / And Earth self-balanced on her center hung" (7.239–42). This curious moment recuperates the animist vision of the universe. What is separated now is not the living from the adverse to life, but different manifestations of the living. What disparts from the matter of the Earth is the matter of the other celestial bodies, which, as Raphael explains to Adam a book later, will be "other suns…/ With their attendant moons…/ Communicating male and female light, / Which two great sexes animate the world" (8.148–51).[55] The Earth itself comes into being, as we saw above, in an alchemical vitalist process as both *in* a womb and *as* itself a womb.

For both Newton and Milton, what is born must die. Following his description of the Earth as "resemble[ing] a great animall," Newton adds that the Earth "according to the condition of all other things living ought to have its times of beginning youth old age & perishing."[56] The entire Earth, not merely substances on earth with their cycles of generation, putrefaction, and regeneration, is subject to old age and death. In early verse for an academic disputation during the 1629 Cambridge Commencement Exercises, Milton wrote in support of the position, as the title for the verses in the 1645 *Poems* has it, "That nature is not subject to old age/*Naturam non pati senium*."[57] Having been assigned the position, most well-known from George Hakewill's *An Apologie of the Power and Providence of God* (1627), against the argument of Godfrey Goodman's in *Fall of Man* (1616) that the world is and has been decaying since the Fall, the young Milton's speaker concludes that the "omnipotent father…poised fate's scales with a pure weight, and commanded each thing to keep a perpetual movement within a supreme order."[58] The world goes on with undiminished vigor. "Phoebus," for example, "gleams with the flower of eternal youth / *Floridus aeternum Phoebus iuvenile*

[55] For the importance of these lines for Milton's influence on later philosophy, see Sanford Budick, *Kant and Milton* (Cambridge, MA: Harvard University Press, 2010), 18–41, and his *Sublime Reciprocity in Milton, Kant and Wordsworth* (Edinburgh: Edinburgh University Press, 2025), 88–89 and *passim*.

[56] Newton, "Of natures obvious laws," 3^v.

[57] For the dating in 1629, Milton's twenty-first year, see *The Complete Works of John Milton, Vol. 3: The Shorter Poems*, eds. Barbara Kiefer Lewalski and Estelle Haan (Oxford: Oxford University Press, 2012), xxxvi, n. 15.

[58] "Pater omnipotens…/…certo…peregit/Pondere fatorum lances, atque ordine summon/Singula perpetuum iussit servare tenorem" (lines 33–36), *MLM* 217.

coruscat" (l. 41). The Book of Revelation (and 2 Peter 3:10), however, forecloses the possibility of that order and youth being eternal:

> Sic denique in aevum
> Ibit cunctarum series iustissima rerum,
> Donec flamma orbem populabitur ultima, late
> Circumplexa polos, et vasti culmina caeli,
> Ingentique rogo flagrabit machina mundi. (65–69)

So, then, the very just course of all things will continue forever, until the last fire devastates the world, broadly enveloping the poles and the summits of vast heaven, and the fabric of the universe blazes on a huge pyre.

When Milton returns, unconstrained by a disputation assignment, to the end of history several decades later in *Paradise Lost*, he implies the aged weariness of the world and expresses the final burning in alchemical terms:

> so shall the world go on,
> To good malignant, to bad men benign,
> Under her own weight groaning till the day
> Appear of respiration to the just,
> And vengeance to the wicked, at return
> Of him so lately promised to thy aid
> The woman's seed, obscurely then foretold,
> ....
> to dissolve
> Satan with his perverted world, then raise
> From the conflagrant mass, purged and refined,
> New heav'ns, new earth, ages of endless date
> Founded in righteousness and peace and love. (12.537–50)

The world's "groaning" expresses not only the travails of the just at the hands of the wicked but also the weariness and pains of an aged world. The fire of 2 Peter 3:10 ("the earth also and the works that are therein shall be burned up") is now conceived as an alchemical fire that will "purge" and "refine." The "dissolving" of the old world recalls the reduction of substances to a dark chaos from which, alchemists such as Newton argued, new bodies are formed. The cyclical and alchemical breathing of Raphael's terrestrial/cosmic animal, the "humid exhalations" rising from the ocean to the sun (*Paradise Lost* 5.423–25), and of Newton's Earth's "draw[ing] in aethereal breath,"[59] will have their crowning moment with the "respiration of the just" with the creation of a world "purged and refined" at

[59] Newton, "Of natures obvious laws," 3ᵛ.

the end of time. What happens continually to bodies in creation happens twice in the linear Christian myth, as the world was formed once from a chaos, and it will be eventually be reduced to a chaos from which a new and final world is made, the "New heav'ns, new earth" of Revelation 21:1.

Medical Alchemy and Living Matter

For both Milton and Newton, the living Earth and macrocosm operate according to alchemical principles. The Earth is an organism, breathing in and breathing out nutritious vapors, and those vapors rise to the heavens where they nourish the sun and the other stars. As the alchemically respiring Earth is more than merely analogous to the individual living animal body, so the processes of individual animal (and human) bodies are explicable in terms of alchemy. The interests of Milton and Newton embrace iatrochemistry, or medical alchemy, as well as chrysopoeia, or the transformation of metals. There is substantial evidence of Newton's keen interest in medical alchemy.[60] And medical alchemy plays a particularly significant role in Milton's depiction of the human makeup and the relation of bodies and thinking.

Newton's interest in medical alchemy is suggested by his transcribing this sentiment in the preface to an alchemical manuscript: "This Philosophy is not of that kind which tendeth to vanity & deceipt but rather to profit & to edification inducing first the knowledg of God & secondly the way to find out true medicines in the creatures."[61] Newton shows an early interest in medical alchemy in the English and Latin manuscripts found on the same folios and discussed above, "Natures obvious laws" and "Humores minerales." In "Natures obvious laws," he asserts that "the two Elixirs [i.e., potable versions of the philosophers' stone] are the most amicable & universall medicine to all beings what ever."[62] He speculates "whither the praeparation of salts by putrefaction if it can bee truly attained would not prove a noble way of Physick."[63] The key evidence for Newton of the continuity of animal and mineral kingdoms (and thus of the vegetable kingdom that lies between) is the demonstrable salutary effect of alchemical treatments on human beings:

> That nothing has so great powers on animalls as mineralls witnes not only the Alkahest to destroy & the Elixir to conserve but their operations in common

[60] See William Newman, "Newton's Reputation as an Alchemist and the Tradition of Chymiatria," in *Reading Newton in Early Modern Europe*, ed. Elizabethanne Boran and Mordechai Feingold (Leiden: Brill, 2017), 313–29.

[61] Newton, "Manna," Keynes Ms. 33, 5^{v}; http://webapp1.dlib.indiana.edu/newton/mss/norm/ALCH00022, accessed April 18, 2019.

[62] Newton, "Of natures obvious laws," 1^{r}.

[63] Newton, "Of natures obvious laws," 2^{v}.

> chymicall physic....They therefore unite with our bodys & become parte of which they could not doe if they had not a principle of vegetation in them & that of the same disposion [sic] with the matter of the rest of our body.[64]

The processes of the vegetation and putrefaction of minerals are continuous with organic processes of the human bodies, providing further proof of the generation of all bodies across animal, vegetable, and mineral kingdoms from one first matter according to the same alchemical laws.

From early in his career, Milton drew from the medical thinking of his day for images. Kester Svendsen catalogues Milton's references to medicine, noting their particular frequency in the prose works.[65] Svendsen for the most part identifies Galenic humoral medical references and references to Renaissance faculty psychology. But he stresses also Milton's "early and constant interest in medical lore," as well as his endorsing an opinion common to medical thinking his day, the "identity between man the microcosm and the universe the macrocosm" (159, 161). For both reasons, one might expect Milton to be alert to developments in medical alchemy. Moreover, his interested curiosity about the causes and possible cures for his blindness and gout make it likely that he discussed these matters with Paget, who collaborated with the medical alchemist Glisson.

The connection to Glisson is particularly relevant, as Nicholls has demonstrated.[66] Glisson is remembered both for his work on rickets, *De rachitude* (1650), and on the anatomy of the liver, *Anatomia hepatis* (1654). As noted in the previous chapter, Glisson like William Harvey came to see life as neither a product of incorporeal spirit nor of highly organized matter; instead, he viewed vital matter as giving rise to organized bodies. Influenced by Harvey and working on the question of tissue irritability, Glisson developed a theory of energetic matter possessing life and perception in itself. This perception differed from animal sensibility in requiring neither organs nor nerves. In his 1672 *Tractatus de natura substantiae energetica, seu de vita naturae*, Glisson, according to Ann Thomson,

> denied that matter was dead and lifeless, claiming that it possessed an energetic nature and not only the capacity to move itself but also, most importantly, 'natural perception'. The different vegetative, animal, or sensitive functions were accounted for by differences in organization, but all bodies have natural life. Glisson thus developed a monistic view of substance as energy, a conception of *biusia* which went against the prevailing mechanistic accounts and undermined substance dualism.[67]

[64] Newton, "Of natures obvious laws," 1^r–1^v.

[65] Kester Svendsen, "Milton and Medical Lore," *Bulletin of the History of Medicine* 13 (1943): 158–84.

[66] For groundbreaking work on the importance of Glisson for an understanding of Milton's thinking about matter, see also Rogers, *Matter of Revolution*, 104–22.

[67] Ann Thomson, *Bodies of Thought: Science, Religion, and the Soul in the Early Enlightenment* (Oxford: Oxford University Press, 2008), 70.

This conception of matter as alive reflects the views of the medical alchemist Jan Baptist van Helmont, and alchemical conceptions play an important role in Glisson's work.[68] Glisson's key departure from the Harveyan model of blood as homogenous matter is the position that blood is a mixed body of "interacting chymical elements."[69]

Nicholls, drawing on the writings of Glisson and of Glisson scholar Guido Giglione, has argued persuasively that Glisson's work on vital matter offers an illuminating context for Milton's picture of the living "one first matter" in *Paradise Lost*. Of particular relevance is Glisson's understanding of the volatilization and sublimation of the corporeal spirits of the blood, described here by Giglione:

> Vital heat is caused and maintained by a chemical reaction occurring continuously in the blood through its circulation and volatilization. At each cardiac contraction, the blood in the heart is kindled as it were and becomes arterial and vital. The saline and sulphureous components of the blood are extremely active and they are the main responsible for its volatilization. 'When they [these components] rise to such a high rank, they gain the "honor" of being called vital spirits.'[70]

In Glisson's theory, the vital, active blood exalts and sublimes itself into purer and more refined versions of itself. According to Nicholls, in both Glisson and Milton, "the upward, transformative urge to ever more perfect formation and ever purer and more powerful spirit is part of a continuum linking the natural world and the human body.... [T]he new chymical *anatomia animata* offers Milton a scientifically progressive version of transformative material spirit."[71] The alchemical sublimation and rarefaction of Glissonian spirits from the vital blood is a credible model for Milton's depiction of the subliming of vegetable food to "vital spirits..., to animal, / To intellectual, [which] give both life and sense" and ultimately reason (*PL* 5.483–88). Glisson's hylozoist matter could move itself and exhibit both perception and appetite, by what was known as a similary (i.e., non- or pre-organic but living and sympathetic) process. A key similary process is the attraction of like matter to like, which, as Glisson concluded, depended in his view on matter's primitive perception and appetite and on the ability to move itself. In Glisson's theory, the powers of perception and appetite accounted for "the ability of matter

[68] Thomson, *Bodies of Thought*, 71.

[69] Nicholls, "Body Out of Spirit," 130.

[70] Guido Giglione, "The Genesis of Francis Glisson's Philosophy of Life" (Ph.D. thesis, John Hopkins University, 2002), 78, quoted in Nicholls, "Body Out of Spirit," 130. In the quoted passage, Giglione paraphrases and quotes Glisson's 1654 *Anatomia hepatis*, 325, 340–41, and Glisson's "Medical Lectures at the College of Physicians," MS Sloane 3309, ff. 370r–v.

[71] Nicholls, "Body Out of Spirit," 131.

to form itself."[72] The "'conglob[ing of] / Like things to like' in Milton's creation account appears to be a similary process."[73] According to John Henry, the "powers of matter enabled Glisson to explain how parts of matter could organize and 'improve themselves' to produce sentient animals and man."[74]

Glisson's understanding of the term "spirits," described here by Nicholls, is particularly useful in dispelling the idea that the word in Milton and Newton's time necessarily denoted immateriality:

> "The word spirit," Glisson writes, "insofar as it is attributed to the bodies, has different meanings." It can generically refer to any kind of body which has been rarefied to the point of becoming volatile (in this respect "wind, air, breath and exhalations in general" were rightly viewed by the ancient authors as sorts of spirit). It can also mean "any body that is subtle, active, and very penetrating." Finally spirit can also mean "that element which, after a due process of fermentation, but not before, strives upwards spontaneously, and becomes volatile." This is "the most precise meaning of the word *spirit*, in that it is understood as an elementary part of a compound. The chemists call it *mercury*.'"[75]

The identification of "mercury" as the "most precise meaning of the word spirit" confirms the provenance of Glisson's vocabulary (and conceptual framework) in medical alchemy. Glisson draws a picture of the progressive sublimation of grosser bodies into purer and purer spirit, precisely parallel to Milton's vision of matter "by gradual scale sublimed" from "flowers and their fruit" to "intellectual spirits." As in the case of the sun supping with the ocean, what can appear to be poetic fancy without the knowledge of the importance of vitalism in seventeenth-century English natural philosophy in fact reflects contemporary natural philosophical developments.

This hylozoist understanding of the life of matter was, as we saw in Chapter 1, the nightmare feared by Ralph Cudworth and Henry More. They viewed the belief that matter is itself alive and self-moving, not needing an incorporeal God to invest it with a self-active incorporeal spirit to account for motion and life, as a short step, if a step at all, from atheism. They worried about vitalist medical alchemy as a particularly dangerous threat. More, speaking of himself in the third person, attacks Glisson in his annotations to Joseph Glanvill's *Lux Orientalis*:

[72] Nicholls, "Body Out of Spirit," 141. [73] Nicholls, "Body Out of Spirit," 140.

[74] John Henry, "The Matter of Souls: Medical Theory and Theology in Seventeenth-Century England," in *The Medical Revolution of the Seventeenth Century*, ed. R. K. French and Andrew Wear (Cambridge: Cambridge University Press, 1989), 111.

[75] Nicholls, "Body Out of Spirit," 138; Nicholls quotes Glisson's *Anatomia Hepatis*, 322, by way of Giglione's "Glisson's Philosophy," 124. As the notes indicate, this section on Glisson's philosophy as analogue for Milton's understanding of living matter and the sublimation of corporal spirits is drawn from Nicholls, who is indebted to Giglione for his work on Glisson.

> For that *Hypothesis*, if it were true, were as safe, if not a safer Refuge for *Atheists*, than the mere Mechanick Philosophie is: . . . the *Atheists* now-a-daies explode the Mechanick Philosophie as not being for their turn, and betake themselves wholly to such an Hypothesis as Dr. *Glissons Vita Naturae*. But, God be thanked, Dr. *H. More* in the fore-cited place has perfectly routed that fond and foul Hypothesis of Dr. *Glisson*.[76]

Cudworth agrees. For him, more insidious than the "atomick atheist," who believes that life somehow emerges from organization of particles of matter, is the "hylozoick atheist":

> Indeed the *Hylozoick Atheists*, being Sensible somewhat of this Inconvenience, of making all *Life* and *Vnderstanding* Out of *Nothing*, and that there must of Necessity be some *Fundamental Life* and *Perception*, which is not *Accidental* but *Substantial*, and which was never *Generated* and cannot be *Corrupted*; have therefore attributed a kind of *Life* and *Perception* to all Matter as such. . . .
>
> These Hylozoists therefore, to avoid a Deity, Suppose every Atom of Sensless Matter, to have been from all Eternity, Infallibly Omniscient, that is, to know all things without either Errour or Ignorance, and to have a Knowledge before Sense.[77]

Like More, Cudworth singles out Glisson as an advocate of hylozoick atheism, observing that the "Point (That all Life, is not a meer *Accident*, but that there is Life Substantial) hath been of late with much Reason and Judgment, insisted upon, and Urged by the Writer *Of the Life of Nature*."[78] The fears of Cudworth and More can be measured by the exaggerated mischaracterization of Glisson's claims for matter's rudimentary perceptivity.

Although More and Cudworth saw atheism hiding behind the alchemical vitalism of contemporaries such as Glisson, others, as outlined in the previous chapter, viewed vitalist conceptions as supportive of belief in God. Newton and Milton were not alone as adamantly theistic vitalist materialists. Noel Sugimura, in her challenge to arguments for Milton's monist vitalism, suggests that scholars maintaining Milton's monism fall into the "error" that More and Cudworth

[76] Henry More, *Annotations upon the Two foregoing Treatises, Lux Orientalis, or, An Enquiry into the Opinion of the Eastern Sages Concerning the Prae-existence of Souls; and the Discourse of Truth* (London, 1682), 192–93; Early English Books Online, ProQuest doc. 2264208413, accessed April 12, 2022. Thomson (*Bodies of Thought*, 76–77) quotes the first part of this passage. The "fore-cited place" is More's anti-Spinozan *Epistola altera V. C.*

[77] Ralph Cudworth, *The True Intellectual System of the Universe* (London, 1678), 755, 870.

[78] Cudworth, *True Intellectual System of the Universe*, 839. As is his practice, Cudworth does not mention contemporaries by name, either allies such as More or those he attacks, most notably Hobbes; the reference here is to Glisson's *Tractatus de natura substantiae energetica, seu de vita naturae*. This passage is quoted by Thomson, *Bodies of Thought*, 78.

diagnosed in their contemporaries. She points to Raphael's creation account, in which "the Spirit of God outspread, / And vital virtue infused, and vital warmth / Throughout the fluid mass" (7.235–37), and notes that

> as soon as we claim that this "Spirit of God" is "Milton's name primarily for the energetic principle behind most natural processes," we run the risk of altering the poem's theology in radical ways: God's actions are immediately naturalized; ideas pertaining to divine conservation (where the *conservation* of an effect depends on God per se and immediately) as well as to *divinus concursus*, or divine concurrence (where the *production* of an effect requires God's cooperation) are effectively removed.[79]

This logic is disputable rather than demonstrative, and indeed it was disputed vigorously in Milton's time and not just by vitalist natural philosophers and experimentalists. The theologian and controversialist Richard Baxter, whom we met in the previous chapter, writes, "I doubt not, *materia qua materia* or yet *qua mere modificata* hath no Life: But that it is *uncapable* of it; and that Almighty God cannot make *perceptive living Matter* and that by *informing* it without mixture, I cannot prove, nor I think you: Where is the Contradiction that makes it impossible?"[80] Baxter does not commit to the position that there is living, perceptive matter "without mixture," that is, without being conjoined with a separate, incorporeal substance, but he thinks it possible, given divine omnipotence. Various figures in the seventeenth century, including Milton, Newton, Cavendish, and Glisson, believed that God did indeed invest corporeal substance with life, and that in doing so he exercised his creative power as efficient cause in the service of final causes, his ends in creating. Newton is particularly adamant on God's exercising final cause and on divine concurrence with efficient causation in created beings, points that Sugimura suggests are incompatible with vitalist conceptions of matter.

Alchemy and Theology

Why were both Milton and Newton attracted to vitalist conceptions of matter and vitalist alchemy in particular? As I have begun to suggest in the preceding chapter, the reasons are theological. Ironically, given More's and Cudworth's fears in their time and Sugimura's argument in our own, their metaphysical positions

[79] Noel Sugimura, "Milton and Matter," Oxford Handbook Topics in Literature (online edition, Oxford Academic, 16 December 2013, 12); https://doi.org/10.1093/oxfordhb/9780199935338.013.11; she quotes Rogers, *Matter of Revolution*, 132.

[80] Richard Baxter, *Of the Nature of Spirits, Especially Mans Soul* (London, 1682), 28–29, quoted in Thomson, *Bodies of Thought*, 76.

derive from the adamant intent of each to defend his understanding of God, proving divine justice in Milton's case and divine omnipotence in Newton's. For Newton, alchemical vitalism was evidence of God's original and continuing creative power; for Milton, it underlined the freedom with which God endowed the world and its creatures. In his natural philosophical work Newton was pursuing theology by other means. In his vitalist understanding of the one first matter, Milton was working out his theodicial imperatives.

Newton approached his diverse and apparently unconnected intellectual pursuits as pieces of one puzzle, all in the service of the explication of the dominion and power of the one God. His alchemical pursuits were no exception, for a reason articulated in an alchemical manuscript that he transcribes:

> For Alchemy tradeth not with metalls as ignorant vulgars think, which error hath made them distrust that noble science; but she hath also material veins of whose nature God created Handmaids to conceive & bring forth his creatures. for it is proper to God alone to create something of nothing, but its natures task to form that which he hath [created.] Wherefore if the fool which hath said in his heart that their [sic] is no God, will put away the mists of ignorance…& behold the power & wisdom of God in his creatures manifested more particularly by the Art of Alchymy imitating Nature in separating from substance, be it vegetable mineral or animal, these three, salt sulphur & mercury, he shall by that mystery as in a glass discern the holy & most glorious trinity in the unity of one hypostasis divine.[81]

Newton may not have endorsed the affirmation of the Trinity at the end of this passage, but his own alchemical writings echo the rest of the passage. In "Of natures obvious laws" Newton writes, "Of protoplasts that nature can onely nourish, not form them, Thats Gods mechanisme the other natures."[82] Newton's "protoplasts" are the alchemical "Handmaids" in the transcribed manuscript, created by God "to conceive & bring forth his creatures." God created the vital matter that in turn vivifies the world, producing and reproducing living animals, vegetables, and minerals. Newton insists, as if in answer to the concerns of More and Cudworth, that the living principles of matter are created or formed by God; nature can only "nourish" them. In the 1675 "Hypothesis of Light" letter to Oldenburg (as secretary of the Royal Society), Newton speculates on the possibility that all of nature is the result of alchemical transformation of aethereal spirits:

> Perhaps the whole frame of nature may be nothing but various contextures of some certain æthereal spirits, or vapours, condensed as it were by precipitation…;

[81] Newton, "Manna," Keynes Ms. 33, 5^{v}; https://webapp1.dlib.indiana.edu/newton/mss/norm/ALCH00022, accessed January 22, 2024.
[82] Newton, "Of natures obvious laws," 1^{r}.

> and after condensation wrought into various forms; at first by the immediate hand of the Creator; and ever since by the power of nature; which, by virtue of the command, increase and multiply, become a complete imitator of the copies set her by the protoplast.[83]

For Newton the alchemical active principles fell under the umbrella of "natures active laws," which accounted for phenomena not explicable through the interaction of "grosser corpuscles" (i.e., particles), their mere arrangement and rearrangement, which could be accounted for by mechanical laws. Such arrangement and rearrangement of passive particles, which could lead to changes in "sensible qualitys," were for Newton superficial relative to the deeper and active changes that he designated as "vegetative":

> There is therefore besides the sensible changes wrough [sic] in the textures of the grosser matter a more subtile secret & noble way of working in all vegetation which makes its products distinct from all others & the immeadiate seate of thes operations is not the whole bulk of matter, but rather an exceeding subtile & inimaginably small portion of matter diffused through the masse which if it were seperated there would remain but a dead & inactive earth.[84]

The vegetative spirits were those copies of the protoplasts, made by the immediate hand of God in creation. Newton, who viewed natural philosophy as instrumental for the demonstration of the activity of an omnipotent creator, was suspicious of mere mechanism, and of the implications of Cartesianism in particular. In "De gravitatione et aequipondio fluidorum," writes, "I shall venture to dispose of [Descartes'] fictions."[85]

Recall Newton in the *Opticks*: "We cannot say that all Nature is not alive."[86] While parcels of matter can be moved passively by collision into rearrangement into new forms, there is a deeper, subtler, and living process beneath the mechanistic interactions that for thinkers such as Hobbes accounted for all phenomena and for thinkers such as Descartes accounted for all phenomena except thought. According to Betty Jo Teeter Dobbs, "Gone, to Newton's mind, were the inert particles of Cartesian matter resting quiescently together between impacts."[87]

83 Newton, "Hypothesis explaining the properties of light," 250.

84 Newton, "Of natures obvious laws," 5v–6r.

85 Newton, "De Gravitatione et Æquipondio Fluidorum," Cambridge University Library MS Add. 4003, https://www.newtonproject.ox.ac.uk/catalogue/record/THEM00093; transcribed and translated in *Unpublished Scientific Papers of Isaac Newton*, eds A. Rupert Hall and Marie Boas Hall (Cambridge: Cambridge University Press, 1962), 123. The Latin reads, "Figmenta ejus tollere conabor" (92).

86 Query 23 in the "Draft version of Queries to the *Optics*," Cambridge University Library ms. 3970.3, fol 620v, accessed August 18, 2025.

87 Betty Jo Teeter Dobbs, *The Foundations of Newton's Alchemy, or, The Hunting of the Greene Lyon* (Cambridge: Cambridge University Press, 1975), 212.

Newton believed that the continuous action of God, in part through the protoplasts or alchemical handmaids, was necessary to maintain the natural order. In the 1675 "Draft Version of the Queries to the *Opticks*," Newton attempts to refute Descartes' mechanist theory of vortices as explanation for action in the physical world and conservator of the cosmos. These vortices could not continue in motion indefinitely,

> but unless the matter were void of all tenacity & attrition of parts & communication of motion (which is not to be supposed) the motion would constantly decay. Seing [sic] therefore the variety of motion which we see in the world is always decreasing, there is a necessity of conserving & recruiting it by active principles; such as are the cause of gravity by which Planets & Comets keep their motions in their Orbs & all bodies acquire great motion in falling; & the cause of fermentation by which the heart & blood of animals are kept in perpetual motion & heat, the inward parts of the earth are constantly warmed bodies burn & shine, mountains take fire, the caverns of the earth are blown up & the Sun continues violently hot & lucid & warms all things by his light. For we meet with very little motion in the world besides what is visibly oweing to these active principles, & the power of the will.[88]

God's mere establishment of mechanical laws and providing initial motion to particles is insufficient in Newton's eyes to account for the world that we observe. A hypothetical divinely created universal perpetual motion machine would be subject to the same attrition through friction as one a human being might attempt to construct. Moreover, the Cartesian model is an affront to God's continual active presence in the world. Note that in this passage from the "Draft Queries," Newton includes the "active principles" by which God "conserve[es] and recruit[s]" cosmological motions in a list with the active alchemical principles of fermentation and vegetation.

Motion in the world derives, as Newton writes, from "these active principles, & the power of the will." Given that we do not move the stars, planets, and comets, that will must be God's. God moves the universe in a way analogous to the way we move our limbs.[89] The alchemical active principles participate for Newton in a network of active principles through which God continues to act in the world. In the chapter on Arianism I will suggest, following Dobbs, that the Son of

[88] Newton, "Draft Version of Queries to the *Opticks*," 255^{r}–256^{r}.

[89] See the passages quoted from Query 23 of the "Draft version of Queries to the *Optics*," fol. 619^{r}, in Chapter 1, nn. 36 and 41: "Life & will are active Principles by which we move our bodies, & thence arise other laws of motion unknown to us... if there be an universal life & all space be the sensorium of a thinking being who by immediate presence perceives all things in it as that which thinks in us perceives their pictures in the brain: the laws of motion arising from life or will may be of universal extent."

God acts as a kind of orchestrator or even embodiment of alchemical active spirit in creation.[90] We have seen in "Of natures obvious laws" and the "Hypothesis of Light" that nature's actions are guided by the active principles infused by God's protoplasts, and that that imitation is vital and ongoing.

Newton's alchemy was neither mystical nor occult, except in the sense that its processes are practically or temporarily occult, not theoretically occult, a distinction that we have seen Robert Boyle make. Vegetative natural processes were subtler and less immediately visible or grossly observable than mechanical processes, but they nevertheless follow laws that were at least theoretically assignable.

The vitalist Newton who wrote that "We cannot say that all Nature is not alive" shared a motivation that I have in the past attributed to Milton, a theologically inspired response to varieties of mechanism. If Milton's motivation is implicit and betrayed indirectly through what I have argued is the dualistic and Cartesian rhetoric of his devils, Newton is explicit in his concern about the dangers of Cartesianism and of the kind of mechanism espoused by Hobbes. For Newton the impetus behind an alchemical exploration in natural philosophy is the identification and publication of divine creative activity in the world. God works by and through the mediation of active principles. They do not take the place of divine creation of and participation in natural phenomena; they are precisely examples of that participation.

Milton deploys alchemical ideas in his imaging of divine participation in the world. The keynote here is freedom. Moral ascent is inseparable from alchemical sublimation of the body, as moral degradation is inseparable from alchemical descent. Here, as in everything else in Milton's world, freedom is fundamental. As we can choose through the exercise of significant moral freedom to rise or fall, to become more spiritual (i.e., more corporeally tenuous), so matter in itself has rudimentary choice, to adhere to what is similar or to what is refining, to purge itself of impurity or be purged as impurity.

Nicholls and Rogers have demonstrated that Glisson's alchemical model of living matter, with its irritability and natural perception, allows for functions that were otherwise associated either with matter animated by an incorporeal soul or, at the least, complex organized structures of matter. The demonstration of unorganized matter's attraction to similar matter and aversion from dissimilar matter sets the stage for the agency of matter that is evident in Milton's account of creation. This does not, as I have suggested, mean that living matter obviates or negates the creative agency of the divine. Here, again, the more scientifically elaborated model in Newton can point us toward the possible meanings behind Milton's active matter in *Paradise Lost*. Newton's active spirits, or active principles, are infused in matter by a divine creator, just as into "Matter unformed,"

[90] Dobbs, *Janus Face*, 83.

Milton's "Spirit of God.../...vital virtue infused, and vital warmth" (*PL* 7.233, 235–36). In turn these active principles activate latent spirits in apparently inert matter. I return one last time to "Of natures obvious laws":

> That clay which is a great ingredient of this upper crust is nothing but stone poudered. that salts may putrefy & by putrefaction will generate another sort of blackish rotten substance < or > fat substance the most fertile part of this upper crust & the nearest matter out of which vegetables are extracted & into which after death they returne.... they are alone as I may say dead & have noe active principle of vegetation in them till they bee incited to it by other substancees that are in a live & vegetating state of what they are mixed withall & thence it is that where nature is not powerfull enough to incite them to action they on the contrary retard & hinder her working.... But if their latent principle can bee once exerted it shows it selfe more vigorusly.[91]

Here is one more key to the passage that might seem to undermine the claim for Milton's animist materialism, Raphael's comment that the "one first matter" is "Endued with various forms, various degrees / Of substance, and in things that live, of life" (5.472–74). What is this matter not indued with life? Newton suggests that there is a latent principle of life even in the most apparently dead matter, that even here a latent life awaits incitement or kindling, similar to the kindling of inanimate spirits by animate spirits in Bacon's *Novum Organum* II, xl, and *De viis mortis*.[92] For Newton bodies that seem "dead & [to] have no active principle" are in that state only "till they bee incited...by other substances that are in a live & vegetating state." Apparently dead matter is potentially alive, and the principle of life resides not in incorporeal substance, but in matter indued with active principles. In Newton's terms these are Nature's copies of the divine protoplast. Milton's alchemical picture of the progression of progressively more spiritual, vital, and dynamic matter out of apparently inert matter is, like Newton's, a way of tying all matter to a living, and in Milton's case free, God. The black tartareous cold infernal dregs, which have exercised Rogers and others, are best understood as fitting a vitalist alchemical model. Milton, I would propose, is not renouncing his vision of matter as both living and good. Granted that he is not a rigorous natural philosopher, he is following a model of living matter that was available to him, as to Newton, from Glisson, Paget, and vitalist experimentalists of the Royal Society.

Milton and Newton: Matter, God, and Modernity. Stephen M. Fallon, Oxford University Press.
 DOI: 10.1093/9780198988939.003.0003

[91] Newton, "Of natures obvious laws," 2^{v}.

[92] See Chapter 1, n. 56.

PART II
RELIGION AND THEOLOGY

3
Two Prophets

Isaac Newton begins an untitled manuscript on the Book of Revelation in terms that will be familiar to readers of John Milton: "Having searched ~~[& by the grace of God obteined~~ \after/ knowleg in the prophetiq3 scriptures, I have thought my self bound to communicate it \for the benefit/ of others, remembring y^e judgment of him who hid his talent in a napkin."[1] While Newton crosses out the explicit claim that he has "by the grace of God obtained" prophetic knowledge, the rest of the sentence all but explicitly stakes a claim to prophetic status that recalls Milton's, and in the same terms. Newton alludes to the Parable of the Talents (Matt 25:14–30). The mortal judgment on one who hides his talent ("cast ye the unprofitable servant into outer darkness," v. 30) spurs Newton's fear of the consequences for one "who hid his talent in a napkin." Milton returns repeatedly to the parable over decades as he considers his prophetic vocation, most famously in Sonnet 19 on his blindness. There, in his mid-40s, he anxiously entertains the possibility that he will have squandered his gifts:

> When I consider how my light is spent,
> Ere half my days, in this dark world and wide,
> And that one talent which is death to hide,
> Lodged with me useless, though my soul more bent
> To serve therewith my Maker. (Sonnet 19.1–5)

As Dayton Haskin has demonstrated, Milton's "talent" here refers primarily not to poetic gifts but to prophetic knowledge of God's word.[2] As we will see, after the age of the Scriptural prophets the role of the prophet was not to share visions from God but to faithfully and authentically interpret the written Word of God. One's knowledge of the Bible and of doctrine was to be cultivated and shared for the benefit of others. If anything, Milton's references to "*one* talent" is a gesture not of pride in poetic gifts but of humility; he compares himself to the servant who is entrusted with the least by his master, and who ignominiously hides rather than propagates that little. The master's response in the parable is harsh:

[1] Extract from Untitled Treatise on Revelation (section 1.1), Yahuda Ms. 1.1, National Library of Israel, Jerusalem, 1^r; https://www.newtonproject.ox.ac.uk/view/texts/diplomatic/THEM00135, accessed April 14, 2022.

[2] Dayton Haskin, *Milton and the Burden of Interpretation* (Philadelphia: University of Pennsylvania Press, 1994), 29–53.

> Thou wicked and slothful servant, thou knewest that I reap where I sowed not, and gather where I have not strawed: Thou oughtest therefore to have put my money to the exchangers, and then at my coming I should have received mine own with usury. Take therefore the talent from him, and give it unto him which hath ten talents. For unto every one that hath shall be given, and he shall have abundance: but from him that hath not shall be taken away even that which he hath. And cast ye the unprofitable servant into outer darkness: there shall be weeping and gnashing of teeth. (Matt 25:26–30)

The thirty-three-year-old Milton had alluded to the parable the previous decade in 1642, in *The Reason of Church-Government*, in terms that echo the parable's language of investment and return:

> he who hath obtain'd in more then the scantest measure to know any thing distinctly of God, and of his true worship, and what is infallibly good and happy in the state of mans life, what in it selfe evil and miserable, though vulgarly not so esteem'd,...remembering also that God even to a strictnesse requires the improvement of these his entrusted gifts, cannot but sustain a sorer burden of mind, and more pressing then any...the body can labor under; *how and in what manner he shall dispose and employ those summes of knowledge and illumination, which God hath sent him into this world to trade with.* (*CPW* 1:801; my emphasis)

A decade earlier again, in 1633, justifying himself for not taking holy orders at the completion of his studies, Milton insisted that he was not failing in "tymely obedience to that comand in the gospell set out by the terrible seasing of him that hid the talent."[3] Instead, he was taking the time to prepare fully to put the talent to work, comparing himself to the last hired laborer in the parable of the vineyard who received the same pay for one hour's work that others' received for a full day's (Matt 20:1–16): "not taking thought of beeing late so it give advantage to be more fit, for those that were latest lost nothing when the maister of the vineyard came to give each one his hire" (*CPW* 1:320). The strained, counterintuitive deployment of the parable betrays uneasiness about his tardiness in taking up the prophetic task of teaching and interpreting the Word.

Milton's emphasis on patience and deliberate preparation points to qualities that he will share with Newton. For both, however much the content of their prophetic writings differs, prophecy was a matter of not of passive ecstatic vision but of chaste life, diligent preparation, and indefatigable labor.

[3] "Letter to a Friend, 1633" (*CPW* 1.320).

The familiar image of Milton as prophetic poetic voice was carefully cultivated by the poet himself, most notably in the invocations in *Paradise Lost*:

> chiefly thou, O Spirit, that does prefer
> Before all temples th'upright heart and pure
> Instruct me...
> what in me is dark
> Illumine, what is low raise and support. (1.17–19, 22–23)
>
> celestial light
> Shine inward, and the mind through all her powers
> Irradiate, there plant eyes, all mist from hence
> Purge and disperse, that I may see and tell
> Of things invisible to mortal sight. (3.51–55)
>
> Up led by thee
> Into the Heav'n of Heav'ns I have presumed,
> An earthly guest...
> ...
> fail not thou, who thee implores;
> For thou art Heav'nly. (7.12–14, 38–39)
>
> If answerable style I can obtain
> Of my celestial patroness, who deigns
> Her nightly visitation unimplored,
> And dictates to me slumb'ring, or inspires
> Easy my unpremeditated verse. (9.20–24)

Milton allowed himself at least a show of trepidation lest he was substituting for inspired verse his own merely human words ("if all be mine, / Not hers who brings it nightly to my ear" [9.46–47]). His great epic project of filling in the gaps in the paratactic Genesis accounts of creation and the fall of humankind would seem to call for divine guidance, even as his ambition to "justify the ways of God to men" could strike contemporaries as a presumptuous attempt to measure God's justice by human reason. Andrew Marvell worried in his dedicatory poem for the second edition of *Paradise Lost* that Milton "would ruin (for I saw him strong) / The sacred truths to fable and old song."[4] By the middle of his poem, a reassured Marvell stipulates that the sacred truths are safe in Milton's hands: "things divine thou treat'st of in such state / As them preserves, and thee, inviolate" (33–34). The veracity of the prophetic account and the enabling purity of the prophet support each other. The happy conclusion derives from a prophetic stance markedly

[4] Marvell, "On Mr. Milton's *Paradise Lost*," *MLM* 287.

distinct, as we will see, from the *furor* of the pagan prophets. Milton, according to Marvell, has been a careful craftsperson, not a manic, mantic voice: "Thou hast not missed one thought that could be fit, / And all that was improper dost omit" (27–28). Milton's authority is guaranteed both by divine inspiration and by measured and rational speech.

In the narrator's voice, the blind poet compares himself in several puzzling lines not only to pagan prophetic poets but to prophets inhabiting rather than authoring pagan poetry:

> Those other two equaled with me in fate,
> So were I equaled with them in renown,
> Blind Thamyris and blind Maeonides,
> And Tiresias and Phineus prophets old. (3.33–36)

Despite indicating "Those other two equaled with me in fate," Milton gives us, in two pairs, four names who shared his fate of blindness. Thomas Newton in 1749 commented on the anomaly: "The most probable explanation of this passage I conceive to be this. Tho' he mentions *four*, yet there are but *two* whom he particularly desires to resemble, and those he distinguishes both with the epithet *blind* to make the likeness more striking."[5] But Milton extends the circle of prophetic vision beyond the two poets to embrace the two divine seers, as he stakes a claim to prophetic vision not to be circumscribed or diminished as poetic convention. While Milton sees his poetry as part of his prophetic utterance, his prophetic stance is not restricted to his vocation as poet.

Though less well known than Milton's claims to prophetic status, Newton's are, though sometimes implicit, just as emphatic. While Milton's self-representations as a prophet appear in one of the most familiar works of world literature, Newton's appear in multiple unpublished manuscripts, particularly his "Treatise on Revelation," and a seldom-read posthumous book, his *Observations upon the Prophecies of Daniel, and the Apocalypse of St. John* (1733). The meaning of the two prophetic books of the canonical Bible would remain obscure until a time closer to the Second Coming of Christ, at which point, Newton wrote, "some should prophesy out of them in an afflicted and mournful state for a long time, and that but darkly."[6] Newton viewed himself as one of these prophets, seeing his task as confirming that subsequent history has borne out the prophecies of Daniel and Revelation, thus demonstrating God's omnipotence. He labored diligently at the work that he describes in this passage, viewing himself as a prophetic

[5] *Paradise Lost*, with Notes of Various Authors, by Thomas Newton, DD, 7th ed. (London, 1770), annotation to 3.33.

[6] Isaac Newton, *Observations upon the Prophecies of Daniel, and the Apocalypse of St. John* (Dublin, 1733), 248 (II.1); Eighteenth Century Collections Online, Gale doc. GALEICW0117768093, accessed April 20, 2022.

interpreter and confirmer of the prophecies of the canonical Bible, and casting himself all but explicitly as one of the "Children of the Resurrection" who would be returned to life to reign between Christ's Second Coming and the Last Judgment.[7] As Frank Manuel observes, Newton "saw himself as the last of the interpreters of God's will in actions, living on the eve of the fulfilment of the times. In his generation he was the vehicle of God's eternal truth."[8]

Prophets after the Age of Prophecy

What counted as prophecy in the seventeenth century? For both Milton and Newton, one has to ask about the relation between Protestant claims of being "led by the Spirit" and claims that a Jeremiah or Ezekiel or Daniel or John were conduits for direct divine revelation. For Milton, there is an additional question, which I will take up first: what is the relation of the prophetic Christian poet to the ancient poet moved by the *furor poeticus*?

Abraham Cowley's first annotation to his unfinished epic *Davideis* domesticates the *furor poeticus*:

> The custom of beginning all *Poems*, with a *Proposition* of the whole work, and an *Invocation* of some God for his assistance to go through with it, is so solemnly and religiously observed by all the ancient *Poets*, that though I could have found out a better way, I should not (I think) have ventured upon it. But there can be, I believe, none better: and that part, of the *Invocation*, if it became a *Heathen*, is no less *Necessary* for a *Christian Poet*.[9]

Cowley's notes reveal that he views epic invocation as a time-honored, mandatory generic convention rather than a plea for, and claim of, divine inspiration. His insistence here on the measured following of convention contrasts sharply with the kind of ecstatic expressions of the Spirit within castigated by Henry More in his *Enthusiasmus Triumphatus* (1656) and exemplified by, for example, an Anna Trapnel. It contrasts also with Plato's description of the poet as inspired by divine madness. In the *Phaedrus*, Socrates describes to Lysias the madness that is "a valuable gift, when due to divine dispensation."[10] This gift of madness is of

[7] "Prophecies concerning Christs 2nd coming," fol. 3; https://www.newtonproject.ox.ac.uk/view/texts/normalized/THEM00088, accessed August 8, 2025.

[8] Frank Manuel, *The Religion of Isaac Newton* (Oxford: Clarendon Press, 1974), 23.

[9] Abraham Cowley, "Davideis," in *Poems: Miscellanies, The Mistress, Pindarique Odes, Davideis, Verses Written on Several Occasions*, ed. A. R. Waller (Cambridge: Cambridge University Press, 1905), first annotation to Book 1, p. 266.

[10] *Phaedrus* 244c, in *Plato: The Collected Dialogues*, ed. Edith Hamilton and Huntington Cairns (Princeton: Princeton University Press, 1961), 491; further references to Plato's dialogues are taken from this edition and cited in the text by Stephanus numbers.

four varieties: (1) the madness of prophecy in the form of seeing into the future; (2) the madness of prophecy in the form of guidance in the forms of "prayer and worship"; (3) the madness of poetic inspiration; and (4) the madness of love (244c–45c). For Socrates, these varieties of madness are closely intertwined and share a disposition to the divine. The second directly involves the proper worship of the divine; when employed rightly, the madness of poetic inspiration issues in what the Socrates of *The Republic* calls "hymns to the gods and the praises of good men" (607a); and properly understood and properly received, according to the *Phaedrus*, love madness culminates not in physical union but in reunion with the gods; physical consummation in fact thwarts or delays that reunion (255e–56e).

Plato suggests that without the divine gift of madness, without divine inspiration, the aspirant to poetry is powerless: "if any man come to the gates of poetry without the madness of the Muses, persuaded that skill alone will make him a good poet, then shall he and his works of sanity with him be brought to naught by the poetry of madness" (*Phaedrus* 245a). Socrates tells Ion that poets compose "not when in their senses," adding that "a poet is a light and winged thing, and holy, and never able to compose until he has become inspired, and is beside himself, and reason is no longer in him. So long as he has this [i.e., reason] in his possession, no man is able to make poetry or to chant in prophecy" (*Ion* 534a–b). In the same way, the power of the prophetess at Delphi depended on the descent of divine madness: "It was when they were mad that the prophetess at Delphi and the priestesses of Dodona achieved so much for which both states and individuals in Greece are thankful; when sane they did little or nothing" (*Phaedrus* 244b). It is by virtue of divine possession that poets and prophets share the title of *vates* or seer. The frenzied possession of the priestesses made visible a dynamic that Plato also ascribed to the poet—each speaks without comprehension words supplied through inspiration.

Not surprisingly, there was a competing late antique Christian understanding of the nature of prophecy. As presented by Plato, prophetic madness and poetic madness are continuous, varieties of the salutary madness bestowed by the gods, as a result of which one stands outside oneself, speaking or writing extrarationally. Newton's purported view of poetry as "ingenious *nonsense*" may owe something to this tradition. The late antique Christian understanding of the nature of prophecy separated poetry from prophecy and defined the gift of prophecy in explicit contrast with the frenzied madness of sybils and visionary priestesses. Milton would reunite the roles of Socrates' poet and one of his two prophets (the one offering guidance in prayer and worship) minus the madness and with the prophet as redefined in the Christian tradition. Church Fathers characterized the prophetic affect in terms precisely opposite to the madness or *ecstasis* described by Plato's Socrates, viewing the Hebrew and Christian prophets as comprehending the inspired words, rather than delivering them in moments of ecstasy, of standing outside themselves.[11] Origen in "On First Principles" viewed pagan prophets as mouthpieces for Satan rather than God, and viewed the Christian prophet as inspired without madness, "without any mental disturbance."[12] Justin Martyr insisted that the prophet is marked by serenity rather than ecstasy.[13] In a foreshadowing of the idea of Christian liberty—that one is most free when one's will is most in line with God's—the biblical prophets were seen as voicing at once God's words and their own rational understanding. In a related paradox, the prophet was seen both as chosen by God and as meriting that election by a virtuous life. In William Kerrigan's trenchant phrasing, "Though chosen, the free man chose to be so."[14] This is clearly the case with Milton, who represents himself both as one chosen by the grace of God and as meriting his election to prophetic status.[15]

What, then, is it to make a prophetic claim in seventeenth-century England? According to one way of thinking, Jewish prophecy had ended with Malachi four centuries before Christ, and Christian prophecy had ended with the last book of the canonical Greek Testament, the Book of Revelation. The 1646 Westminster Confession of Faith begins with an assertion of the cessation of prophecy understood as the direct revelation of God's words to human beings: "it pleased the Lord, at sundry times, and in divers manners, to reveal Himself, and to declare that His will unto His Church; and afterwards for the better preserving and propagating of the truth,... to commit the same wholly unto writing;... those former ways of God's revealing His will unto His people being now ceased."[16] The canonical prophets received privileged and direct communication from God. At the same time, this did not preclude the enjoyment of spiritual gifts by others, less expressly chosen and authorized by God. In Numbers 11:27–29, Moses endorses the idea that there can be a wider circle of prophets: "there ran a young man, and told Moses, and said, Eldad and Medad do prophesy in the camp. And Joshua... answered and said, My lord Moses, forbid them. And Moses said unto him, Enviest thou for my sake? would God that all the LORD'S people were prophets, and that the LORD would put his spirit upon them!" The possibility that the Spirit could communicate to those outside the small circle of canonical prophets would resonate in the Greek Testament and beyond. The Westminster Confession's position that prophecy ended with the Greek Testament

[11] The following discussion draws from William Kerrigan, *The Prophetic Milton* (Charlottesville: University Press of Virginia, 1974), 23 ff.

[12] Kerrigan, *Prophetic Milton*, 24, 27, citing Origin's *On First Principles*.

[13] Kerrigan, *Prophetic Milton*, 26, citing *The Discourse to the Greeks*, Chapter 5.

[14] Kerrigan, *Prophetic Milton*, 33.

[15] See my *Milton's Peculiar Grace: Self-representation and Authority* (Ithaca: Cornell University Press, 2007), 185–88.

[16] Westminster Confession of Faith (London, 1646), I.1; https://prts.edu/wp-content/uploads/2016/12/Westminster_Confession.pdf, accessed August 18, 2025.

notwithstanding, the Greek Testament testifies to the continuation of spiritual gifts among Christians, as Paul makes clear:

> Now there are diversities of gifts, but the same Spirit.... For to one is given by the Spirit the word of wisdom; to another the word of knowledge by the same Spirit; To another faith by the same Spirit; to another the gifts of healing by the same Spirit; To another the working of miracles*; to another prophecy*; to another discerning of spirits; to another divers kinds of tongues; to another the interpretation of tongues. (1 Cor 12:4, 8–10; my emphasis)

Moses' wish that all the Lord's people might be prophets and Paul's indication of the wide dispersal of the gifts of the Spirit fit well with Reformed emphases on the priesthood of all believers and *sola scriptura*. As each believer was to work out his or her belief in individual and diligent study of the Bible, led by the Spirit, rather than relying on external authority, principles firmly embraced by Milton and Newton, each believer was at least potentially prophetic.

Milton invokes the Reformed imperatives that one work out one's beliefs through prayerful reading and interpretation of Scripture in the introductory epistle to his *De Doctrina Christiana*: "I resolved to rely on the faith or judgement of no one else as to the things of God, but rather to take my religious belief from a faith based exclusively on divine revelation;... perusing and perpending the scripture of God itself with the utmost diligence, to have every single point investigated and understood for myself" (*OCW* 8.1:5). Milton here is squarely in the Reformed tradition. A century earlier, John Calvin claimed that an authentic reading of Scripture is one led by the Spirit: "as God alone is a fit witness of himself in his Word, so also the Word will not find acceptance in men's hearts before it is sealed by the inward testimony of the Spirit. The same Spirit, therefore, who has spoken through the mouths of the prophets must penetrate into our hearts to persuade us that they faithfully proclaimed what had been divinely commanded."[17] Divine inspiration is as necessary to the reading of Scripture as it was to Scripture's composition.

Milton extends that logic in granting priority to the Spirit over the Word, whatever the Protestant insistence on *sola scriptura*. As he argues in *De Doctrina Christiana*'s chapter on Holy Scripture (I.30),

> we have, particularly under the gospel, a twofold scripture: the external scripture of the written word, and the internal one of the holy spirit, which he... has etched on believers' hearts.... [¶] And so today the external authority for our faith, [contained] indeed in the scriptures, is very great, and generally first, temporally; but each person has the internal authority, and likewise the *supreme*

[17] John Calvin, *Institutes of the Christian Religion*, ed. John T. McNeill and trans. Ford Lewis Battles, Library of Christian Classics, 2 vols (Philadelphia: Westminster Press, 1960), I:79 (I.vii.4).

> *and preeminent* one: the spirit itself [*interna verò cuique, adeóque* summa atque suprema, *est ipse spiritus*]. (*OCW* 8.2:810–11; my emphasis)

The superior authority of the spirit licenses Milton to challenge in the treatise more than a millennium of orthodoxy on the Trinity and to mount in the divorce tracts and the treatise a counterintuitive reading overturning Jesus' apparently clear prohibition of divorce in the Gospels.

Newton also testifies to the authorizing power of being led by the Spirit. In his "Untitled Treatise on Revelation," he points to the eyes in the wings of the four beasts of Revelation 4:8 ("And the four beasts had each of them six wings about him; and they were full of eyes within") as symbolizing this prophetic power:

> where the four Beasts are said to be full of eyes before & behind & their wings to be full of eyes within, this denotes them to signify prophetic multitudes.... Now their [double] eyes in their wings because not without but only within denote an internal faculty of seing that is, their intellect & this illuminated not barely by the light of nature but by the Spirit because they are not mere men but servants & worshippers of the living God Apoc 4.8, 9.[18]

Newton adds here that the array of eyes in Revelation confirms that individuals could be "illuminated not barely by the light of nature but by the Spirit" and that this illumination gives rise to "prophetic multitudes." We will see that for Newton as well as Milton a set of characteristics including chaste life, learning, and diligent labor sets the prophetic figures apart not only from "mere men" but also from vulgar prophets.

The work of interpreting Scripture while led by the Spirit is, for Milton and Newton and their contemporaries, the prophetic work of those John Wolleb describes as "extraordinary ministers." He writes in his 1626 *Compendium Theologiae*, a crucial influence on Milton's *De Doctrina Christiana*,

> The extraordinary ones are those men whom God raises up either for establishing a new regime in the church, *or for restoring the old when it had broken down.* Such, under the Old Testament, were the prophets. In the time of the New Testament, John the Baptist, Christ, the apostles, *the prophets* (i.e. *men with the gift of interpreting Scripture*), the evangelists, who were companions and assistants of the apostles, the pastors of churches, and the teachers in charge of schools.[19]

[18] Newton, *Untitled Treatise on Revelation*, fol. 38A^{v}; https://www.newtonproject.ox.ac.uk/view/texts/normalized/THEM00135, accessed January 23, 2024. The normalized text quoted here does not contain the word "double" (in brackets), despite the fact that it appears as an insertion in the diplomatic text. Further quotations are taken from the normalized transcription cited here.

[19] Johannes Wollebius, *Compendium Theologiae Christianae*, XXVI.2, in *Reformed Dogmatics*, ed. and trans. John W. Beardslee III (New York: Oxford University Press, 1965), 142 (emphasis mine).

By listing pastors and interpreters of Scripture alongside Christ, the apostles, and evangelists, Wolleb leaves a space for a continuing tradition of prophecy.

At the beginning of the Reformation Martin Luther claimed that the interpretation of biblical prophecy is itself prophetic: "There are many kinds of prophecy in the church. One is prophecy which interprets the writings of the prophets. Paul speaks of it in 1 Corinthians xii and xiv, and in other places. This is the most necessary kind…, because it teaches the Word of God, lays the foundation of the Church, and defends the faith."[20] Paul, in addition to his affirming the variety of the gifts of the Spirit in 1 Corinthians 12, makes explicit the connection between prophecy and teaching in 1 Corinthians 14:3–4: "But he that prophesieth speaketh unto men to edification, and exhortation, and comfort…. [H]e that prophesieth edifieth the church." Milton takes up this line in the *De Doctrina Christiana* I.29: "Named 'prophet', too, is not only he who foretells future events, but also he who by his outstanding piety and wisdom is equipped for teaching…. Hence the simple gift of teaching, especially public [teaching], is also, under the gospel, called *prophecy*" (*OCW* 8.2:789).[21] The teaching authority, Milton adds a moment later, is a gift of God not of the universities, "For he who gave apostles and prophets, the same also gave pastors and teachers; Universities did not."[22] Newton, as we shall see below, also thought of the teaching of learned exegetes as prophetic.

There is an obvious difficulty with claims that inward illumination by the Spirit authorizes prophetic testimony. Calvin's formulation, as Kerrigan has noted, is circular: "One must have the Spirit to recognize the Spirit, be inspired to recognize inspiration."[23] Where the Roman church insisted on tradition and institutional authority, the belief that the indwelling spirit is the highest authority had an inevitably and inexorably splintering effect on doctrinal unity. The prophetic authority of the individual Christian collided with the political authority of the king or magistrate and the ecclesiastical authority of the clergy.[24]

The problem with the argument for the indwelling Spirit as authorizing the interpretation of Scripture was precisely its inwardness. However much one might be convinced of the direction of the Spirit, that direction could not be immediately evident to others. Conflicting claims by "prophets" could have no obvious adjudication if each person's oracle was inward. The age abounded in

[20] "Preface to the Revelation of St. John (I)," in the *Works of Martin Luther*, ed. Andrew Spaeth et al., vol. 6 (Philadelphia: Holman, 1932), 479–80.

[21] *"Propheta etiam non tantum is qui future praedicit, set is etiam qui eximia pietate et sapientia ad docendum instructus est, nominator….Hinc simplex donum docendi, praesertim public,* prophetia *quoque dicitur sub evangelio"* (*OCW* 8.2:788)

[22] *"Qui enim dedit apostolos, et prophetas, is etiam dedit pastores et doctores; non Academiae"* (*OCW* 8.2:788).

[23] Kerrigan, *Prophetic Milton*, 96.

[24] For how this played out in New England, see Keith Stavely, *Puritan Legacies: Paradise Lost and the New England Tradition, 1630–1890* (Ithaca: Cornell University Press, 1987).

enthusiasts claiming divine inspiration. The emphasis on the Spirit within made it difficult to maintain a religious denomination as a unified body, as diverging "inspired" readings had an atomizing effect. Thomas Hobbes complained that

> after the Bible was translated into English, every man, nay, every boy and wench, that could read English, thought they spoke with God Almighty, and understood what he said....And so the reverence and obedience due to the Reformed Church here, and to the bishops and pastors therein, was cast off; and every man became a judge of religion, and an interpreter of the Scriptures to itself.... [T]his license of interpreting the Scripture was the cause of so many several sects.[25]

Hobbes contemptuously dismissed claims of inspiration. They did not rise even to the level of falsehood; like assertions of the freedom of the will, they were examples of absurd speech: "The second cause of Absurd assertions, I ascribe to the giving of names of *bodies*, to *accidents*...; As they do, that say, *Faith is infused*, or *inspired*; when nothing can be *powred*, or *breathed* into any thing, but body."[26] Inspiration, Hobbes sardonically observed, "is nothing but the blowing into a man some thin and subtile aire, or wind, in such manner as a man filleth a bladder with his breath."[27] Claims of inspiration interposed an imagined private authority between the individual and sanctioned authority: "And thus wee fall again into the fault of taking upon us to Judge of Good and Evill; or to make Judges of it, such private men as pretend to be supernaturally Inspired, to the Dissolution of all Civill Government."[28] The faculty of prophecy, in Hobbes' view, was the imagination: "the Prophets extraordinary in the Old Testament took notice of the Word of God no otherwise, than from their Dreams, or Visions; that is to say, from the imaginations which they had in their sleep, or in an Extasie: which imaginations in every true Prophet were supernaturall; but in false Prophets were either naturall, or feigned."[29] True prophets for Hobbes were those whose imaginations and night visions were supernatural, a status confirmed (circularly?) by their inclusion in canonical Scripture.

Like Hobbes, Baruch Spinoza associated prophecy with imagination. In his *Theologico-Political Treatise*, he argued that "the power of prophecy implies not a peculiarly perfect mind, but a peculiarly vivid imagination," and that, "Inasmuch as imagination is fleeting and inconstant, we find that the power of prophecy did not remain with a prophet for long, nor manifest itself frequently, but was very

25 Thomas Hobbes, *Behemoth: Or the Long Parliament*, ed. from original manuscript by Ferdinand Tönnies (London: Simpkin, Marshall, 1889), 21–22.

26 Thomas Hobbes, *Leviathan*, ed. C. B. Macpherson (Harmondsworth: Penguin, 1968), 114 (Chapter 5).

27 Hobbes, *Leviathan*, 440 (Chapter 34).

28 Hobbes, *Leviathan*, 366 (Chapter 29).

29 Hobbes, *Leviathan*, 461 (Chapter 36).

rare; manifesting itself only in a few men, and in them not often."[30] While Spinoza seems to honor prophets by emphasizing, like Milton and Newton, the rarity of prophetic imagination, he undercuts the honor by insisting that prophets were not superior to others in knowledge. Joshua was no astronomer, so we need not believe that the sun stood still, and the Magi believed that the birth of Christ was manifest by a new star in the East because they "believed in the follies of astrology" (33, 30; Chapter 2). Spinoza singles out Daniel, a prophet immensely important to Newton, as one who imagined his visions "so confusedly and obscurely that afterwards he could make nothing of them" (32; Chapter 2). For Newton, as we shall see, the obscurity of Daniel's prophecies is a function of God's plan to place a veil between his revelation and the eyes of the wicked; for Spinoza, that obscurity is not strategic but instead a function of Daniel's limited intellect and knowledge. Prophecy cannot offer certain knowledge, a fact confirmed by the fact that prophets demand signs from God of the veracity of visions. "In this respect," Spinoza writes, "prophetic knowledge is inferior to natural knowledge, which needs no sign, and in itself implies certitude" (28; Chapter 2).

The certitude or veracity of prophecy is merely moral. One can extract the moral kernel from the vision adapted to the limited intelligence and knowledge of the prophet: "God adapted revelations to the understanding and opinions of the prophets, and that in matters of theory without bearing on charity or morality the prophets could be, and, in fact, were, ignorant" (40; Chapter 2). Spinoza, however, immediately qualifies his recognition of prophets as moral teachers. The moral revelations enshrined in biblical prophecies are themselves marred by the limitations of the moral vision of the prophets and the people to whom they speak. The Hebrew prophets, with their emphasis that God is the God of Israel and that they were a uniquely elect nation, were ignorant of what for Spinoza is a fundamental moral principle, that

> Every man's true happiness and blessedness consist solely in the enjoyment of what is good, not in the pride that he alone is enjoying it, to the exclusion of others. He who thinks himself the more blessed because he is enjoying benefits which others are not, or because he is more blessed or more fortunate than his fellows, is ignorant of true happiness and blessedness, and the joy which he feels is either childish or envious and malicious. (43; Chapter 3)

Moses himself was further limited by his belief that one should love God because of past benefits, hope of future ones, and fear of punishment. Spinoza concludes that the Israelites worshipped a God who "treat[ed] them as parents treat irrational children" (39; Chapter 2).

[30] Benedict de Spinoza, *A Theologico-Political Treatise*, trans. R. H. M. Elwes (New York: Dover, 1951), 19, 25–26 (Chapter 1); further quotations cited by page and chapter number in the text.

Spinoza locates prophecy in the imagination, limits it to a few authorized voices, circumscribes its authority to morality, and then proceeds to dismantle the authority of those voices by demonstrating that they simply reiterate the limited and even childish moral vision of their times and places. What is left is the understanding of virtue and happiness available to reason, which does not require the aid of prophecy.

Hobbes agrees with Spinoza on the superiority of reason to inspiration: "It hath been also commonly taught, *That Faith and Sanctity, are not to be attained by Study and Reason, but by supernaturall Inspiration, or Infusion*, which granted, I see not why any man should render a reason of his Faith; or why every Christian should not be also a Prophet; or why any man should take the Law of his Country, rather than his own Inspiration, for the rule of his action."[31] Hobbes requires rational evidence for the veracity of prophetic utterance. Study and reason will be the warrants of authority, a position that he shares with Spinoza.

Whatever they may have thought of Hobbes and Spinoza in general,[32] both Milton and Newton viewed their own prophetic roles in terms that addressed not only the criticisms of the two thinkers, thought by many of their contemporaries to be atheists, but also the concerns of the orthodox that Enthusiasts would unleash a host of conflicting interpretations of Scripture. They laid out similar programs designed to narrow the gate for authentic prophetic interpreters. Both emphasized a set of requirements for the authentic interpretation of Scripture, and thus of the authenticity of their prophetic roles, including moral purity, a knowledge of biblical languages, indefatigable labor, and submission of their arguments to strict canons of reason.

The Attributes of a Prophet

If Hobbes feared the cacophony of interpretations and proliferation of sects, both Milton and Newton, notwithstanding the latter's reference to "prophetic multitudes" in his "Untitled Treaties on Revelation," emphasized the exclusivity and narrowness of the circle of prophetic speakers. If Moses wishes that "all the Lord's people were prophets," Milton and Newton, inclining toward Isaiah 1:9 ("Except the Lord of hosts had left unto us a very small remnant, we should have been as Sodom"), viewed their membership in a vanishingly small group as a sign of prophetic election. Milton in *Paradise Lost* called for a "fit audience, though few"

[31] Hobbes, *Leviathan*, 366 (Chapter 29) (my emphasis).

[32] We do have evidence of Milton's general impression of Hobbes, from Elizabeth Minshull Milton. John Aubrey, in his "Minutes of the Life of Mr. John Milton," records that "His widow assures me that Mr. Hobbs was not one of his acquaintance; yet her husband did not like him at all: but he would grant him to be a man of great parts, a learned man. Their interests and tenets did run counter to each other," *MLM*, xxx.

(7.31), and speakers fit to address such audiences are even fewer. In the autobiographical preface to the second book of *The Reason of Church-Government*, Milton describes the selection and preparation of the prophetic poet: the ability to compose such "frequent songs" as appear "throughout the law and prophets,...are the inspired guift of God rarely bestow'd, but yet to some (though most abuse) in every Nation: and are of power beside the office of a pulpit, to imbreed and cherish in a great people the seeds of vertu, and publick civility,...to celebrate in glorious and lofty Hymns the throne and equipage of Gods Almightinesse" (*CPW* 1:816–17). God chooses to share gifts with the few, and those few either cultivate or abuse those gifts. As the Father in *Paradise Lost* tells the Son of God that he possesses his name both "by merit" and "birthright" (3.309), Milton imagines himself both chosen prior to merit and as meriting his role. Newton also counts himself the member of a small and "scattered" community of prophets:

> it was revealed to Daniel that the prophesies concerning the last times should be closed up & sealed untill the time of the end: but then the wise should understand, & knowledg should be increased. Dan 12.4, 9, 10....& if so, then it is as certain that the church shall at length attain to the understanding thereof. *I mean not all that call themselves Christians, but a remnant, a few scattered persons which God hath chosen*, such as...can set themselves sincerely & earnestly to search after truth.[33]

Newton insists that the prophets are rare individuals. As in the case of Milton, the rare prophet speaks to the only relatively less rare godly audience:

> I could wish they would consider how contrary it is to God's purpose that the truth of his religion should be as obvious & perspicuous to all men as a mathematical demonstration. Tis enough that it is able to move the assent of those which he hath chosen; & for the rest who are so incredulous, it is just that they should be permitted to dy in their sins. Here then is the wisdom of God, that he hath so framed the Scriptures as to discern between the good and the bad, that they should be demonstration to the one & foolishness to the other.[34]

Having chosen his small remnant of prophets, Newton's God, now a predestinating God, divides the chosen who accept divine truths from the incredulous reprobate. Newton focuses on exclusivity. Rather than the proliferation of sects and doctrines, he foresees a winnowing of true believers from the chaff of the many.

While convinced that authentic interpretation of Scripture required the guidance of the spirit, Milton and Newton were equally suspicious of ungoverned

[33] Newton, "Untitled Treatise on Revelation," 1r (my emphasis), accessed May 4, 2022.
[34] Newton, "Untitled Treatise on Revelation," 19r.

imagination. Newton complained that heresies (including doctrines that were widely considered orthodox) arose from undisciplined readings based on imagination:

> He that without better grounds then his private opinion or the opinion of any human authority whatsoever shall turn scripture from the plain meaning to an Allegory or to any other less naturall sense declares thereby that he reposes more trust in his own imaginations or in that human authority then in the Scripture. And therefore the opinion of such men how numerous soever they be, is not to be regarded....And this hath been the door through which all Heresies have crept in & turned out the ancient faith.[35]

Without discipline, one wanders from the clear sense of Scripture, and Newton describes the result in terms similar to those with which Hobbes has excoriated vulgar prophets: "Too much liberty in this kind savours of a luxuriant ungovernable fansy and borders on enthusiasm."[36]

In place of ungoverned imagination or fancy Milton specifies in *De Doctrina Christiana* I.30 the "the right method of interpreting the scriptures": "expertise in languages; consideration of sources; attention to [overall] purpose; distinction between literal and figurative language; examination of causes, circumstances, preceding and following events; comparison of texts with other texts; also the 'analogia fidei' [overarching sense of scripture]—[each of these] must everywhere be looked at" (*OCW* 8.2:803; bracketed phrases in the cited text).[37] These are all criteria shared by Newton, who, like Milton, could read Greek and also, with the help of a dictionary, Hebrew. Newton, as we will see, labored to codify the figurative language of the prophetic books. And he also shared Milton's emphasis on deciphering the meaning of possibly ambiguous verses by comparing places in the Bible.

As Haskin has shown, Milton relied on methodical comparison of one biblical verse with another.[38] The figure of Mary in *Paradise Regained*, surprisingly prominent in a Protestant poem, becomes a model for the kind of reading performed by her Son in the poem and by Milton in his exegetical work. A speech that begins in anguished uncertainty about her missing son ends with calm resolve won by careful collation of the parts of prophecy: "My heart hath been a storehouse long of things / And sayings laid up, portending strange events" (2.103–4). Milton in

[35] Newton, "Untitled Treatise on Revelation," 12^r.
[36] Newton, "Untitled Treatise on Revelation," 12^r.
[37] "*Ratio rectè interpretandi scripturas util[ius]*| *quidem à Theologis traditur, quàm diligentius aut fidelius observatur; linguarum peritia; fontium inspectio; scopi animadversio; locutionis propriae et figuratae distinctio; causarum, circumstantiarum, antecedentium, consequentium consideratio; locorum cum aliis locis comparatio; fidei quoque analogia ubique spectanda est*" (*OCW* 8.2:802).
[38] Haskin, *Milton and the Burden of Interpretation*, Chapter 5 "Conferring Places with Mary, Seeking Closure with Manoa," 118–46.

the introductory letter to *De Doctrina Christiana* promotes the disciplined comparison of biblical place with place. He describes years spent reading Scripture and arranging disparate verses into topical tables, all preparatory to the formulation of arguments that would provide a reasoned defense of his faith (*OCW* 8.1:5–7). He emphasizes, defiantly rather than apologetically, that unlike the treatises of other exegetes, his pages will "overflow with scriptural [verses]..., and that as little room as possible be left for my own words" (*OCW* 8.1:9).[39] The gathering and comparing of verses becomes a protection against the construction of specious arguments based on isolated passages. A prime case of this for Milton, and a case that for Haskin was pivotal in turning Milton to this method of exegesis, is Milton's reconciliation of the apparent contradiction between Moses' sanctioning (Deut 24:1) and Jesus's prohibition (Matt 5:31–32) of divorce.[40] In *The Doctrine and Discipline of Divorce*, Milton cites Matthew 5:18, where Jesus insists that "he came not to abrogate from the Law *one jot or one tittle*" (*CPW* 2:283), to argue that the apparent overturning of Moses' permission to divorce later in the same chapter would open Jesus to the charge of contradiction: he would be abrogating much more than a jot or tittle, and he would be making Gospel teaching more rigorous and burdensome than Mosaic law, despite his reassurance that "his yoke is easie and his burden light" (Matt 11:30; *CPW* 2:240). Instead, Milton maintains that Jesus in Matthew 5:31–32 speaks to the hardness of heart of his pharisaical questioners. True marriages are indissoluble, but unions of incompatible partners are not true marriages, as they fail to fulfil the purpose in the institution of marriage, as is clear, Milton argues, from "Genesis 2.18, where God announces that purpose of God's provision of marriage: 'It is not good that the man should be alone; I will make him an help meet for him'" (*CPW* 2:326–29, 245–46; II.16 and I.2).

Newton similarly insisted that proper interpretation of the Bible required careful collation of verses. In his manuscript on Revelation, he lays out his method: "First I shall lay down certain general Rules of Interpretation, the consideration of which may prepare the judgment of the Reader & inable him to know when an interpretation is genuine & of two interpretations which is the best."[41] A crucial step, which he performs repeatedly in his obsessive hermeneutical manuscripts, is explicating prophetic language, mainly by translating (figurative) natural into

[39] For a careful and illuminating analysis of where, how, and why Milton's does not always follow this self-imposed imperative, see Jason Kerr, *Milton's Theological Process: Reading "De Doctrina Christiana" and "Paradise Lost"* (Oxford: Oxford University Press, 2024).

[40] The clashing verses are Moses' "When a man hath taken a wife, and married her, and it come to pass that she find no favour in his eyes, because he hath found some uncleanness in her: then let him write her a bill of divorcement, and give it in her hand, and send her out of his house" (Deut 24:1) and Jesus' "It hath been said, Whosoever shall put away his wife, let him give her a writing of divorcement: But I say unto you, That whosoever shall put away his wife, saving for the cause of fornication, causeth her to commit adultery" (Matt 5:31–32).

[41] Newton, "Untitled Treatise on Revelation," 10^{r}.

(literal) politico-historical terms. Then, he writes, "I compare the parts of the Apocalyps one with another & digest them into order by those internal characters which the Holy-ghost hath for this end imprest upon them. And this I do by drawing up the substance of the Prophesy into Propositions, & subjoyning the reasons for the truth of every Proposition."[42] For Newton that "substance" was historical, to confirm that the prophecies of Daniel and Revelation accurately predicted subsequent historical events. Newton was as exhaustive and meticulous in his reading of the prophetic Books of Daniel and Revelation as he was with his "Chronology of the Ancient Kingdoms Amended," in which he demonstrated the reconcilability of biblical and Gentile histories (and the temporal priority of biblical Egypt as the first great kingdom).

Milton and Newton had similar ideas about the criteria for admission to the rarefied company of divinely authorized prophets, beginning with chastity and sexual purity. Milton describes the rigorously ascetic life of the epic and prophetic poet in his Elegy 6, written shortly after his twenty-first birthday to his friend Charles Diodati,

> he who tells of wars, and of heaven under the rule of Jove in his maturity, and reverent heroes and semidivine leaders, and sings now of the sacred deliberations of the supreme gods, now of the deep realm where the fierce dog barks.... Let him live sparingly, like the master of Samos, and let plants provide him with harmless food; let the clearest water stand nearby in a beechwood vessel, and let him drink sober drafts from a pure spring. Add to this a youth free of crime and chaste, and strict morals, and a hand free from stain. (Elegy 6, 55–64)[43]

So lived Tiresias, Calchas, Orpheus, and Homer (67–73), and living like this will allow the young poet to join that prophetic line. Here, as he would decades later in the invocation to Book 3 of *Paradise Lost*, Milton links two inspired poets with two seers.

Milton linked sexual purity with visionary power explicitly in the *Mask at Ludlow*, *An Apology for Smectymnuus*, *Animadversions*, and the *Reason of Church-Government*.[44] Before his marriage in 1642, Milton emphasized virginity

[42] Newton, "Untitled Treatise on Revelation," 10r.

[43]
qui bella refert, et adulto sub Iove caelum,
Heroasque pios, semideosque duces,
Et nunc sancta canit superum consulta deorum,
Nunc latrata fero regna profunda cane.
.....
Stet prope fagineo pellucida lympha catillo,
Sobriaque e puro pocula fonte bibat.
Additur huic scelerisque vacans et casta iuventus,
Et rigidi mores, et sine labe manus. (ll. 55–64)

[44] He may have been influenced by the argument in Plato's *Phaedrus* that, as noted above, sexual consummation thwarts reunion with the divine (255e–56e).

as essential to visionary power. At the conclusion of his *Epitaphium Damonis*, Milton congratulates his deceased friend Diodati, "Because…a youth without stain pleased you, because the joy of a marriage bed was never tasted, lo, virginal honors are reserved for you" (212–14),[45] namely presence at the celestial wedding feast promised in Revelation 14:1–5. The Lady in the *Mask*, with whom Milton, "the Lady of Christs," in some ways identifies, points in the context of the power of sacred utterance to "the sun-clad power of Chastity" and testifies to the prophetic power available to those dedicated to the "sage / And serious doctrine of Virginity" (782, 786–87). She challenges the tempter Comus:

> Thou art not fit to hear thyself convinced;
> Yet should I try, the uncontrollèd worth
> Of this pure cause would kindle my rapt spirits
> To such a flame of sacred vehemence
> That dumb things would be moved to sympathize,
> And the brute earth would lend her nerves and shake,
> Till all thy magic structures reared so high
> Were shattered into heaps o'er thy false head. (792–99)

In replying, Comus himself testifies to the prophetic power: "She fables not, I feel that I do fear / Her words set off by some superior power" (800–801). As Milton eventually contemplated marriage, he expanded his focus to embrace married chastity. In the *Apology for Smectymnuus*, published in 1642, weeks before his meeting and marrying Mary Powell, his first wife, Milton links virginity and married chastity in celebrating chiastically "the honour and chastity of Virgin or Matron" (*CPW* 1:891). Milton now claims that the reward given "to those who were not defil'd with women" in Revelation, with which he ended his elegy to Diodati, does not require virginity and that defilement "doubtlesse meanes fornication: For mariage must not be call'd a defilement" (*CPW* 1:892–93).

Newton like Milton saw a link between access to truth and personal purity. John Conduitt, who married Newton's half-niece and succeeded Newton as Master of the Mint, describes Newton's perspective on the relation of personal virtue and philosophy:

> the true Philosopher must have temperance justice & a firmness of mind be a lover of truth avoid pleasures & despise riches *break loose as much as is possible from that bond & subjections which the soul is in to the body,* not be afraid of poverty or ignominy or calumny in the cause of justice & truth do good even to

[45] *Quod tibi…sine labe iuventus*
Grata fuit, quod nulla tori libata voluptas,
En etiam tibi virginei servantur honores. (212–14)

enemies, to have no thought but how to die well, & for that purpose to give up every thing & even your self—Sir I[saac] was a Philosopher according to this definition as much in his morals as his writings he joined morality to Philosophy.[46]

If Milton embraced married chastity, Newton as far as we know was a lifelong virgin.[47] Newton, however, joined Milton in condemning enforced clerical celibacy as a Roman Catholic aberration. Newton insisted that the psychological effect of insistence on celibacy was to stir up monks' lustful thoughts:

> it was a general complaint among them that upon their entring into the profession of a Monastick life they found themselves more tempted in the flesh then before.... The reason they gave of it was that the devil tempted them most who were most his enemies: but the true reason was partly that the desire was inflamed by prohibition of lawfull marriage, & partly that the profession of chastity & daily fasting on that account put them perpetually in mind of what they strove against, & their idle lives gave liberty to their thoughts to follow their inclinations.[48]

Milton emphasized occupation and healthy diet to counter lustful thoughts (*Doctrine and Discipline*, *CPW* 2:881). Both Milton and Newton could be morally censorious and demanding of themselves and others, and both emphasized sexual purity as essential for the prophet. In the "Prophecies concerning Christs 2nd coming," Newton includes, as Milton did initially, virginity as a qualification for belonging to this select group: "The children of this world marry & are given in marriage, but they which shall be accounted worthy to obtein that world, & the resurrection from the dead, neither marry nor are given in marriage, neither can they die any more for they are equall unto the angels & are the children of God being the children of the resurrection."[49]

If Milton and Newton agreed on the importance of moral purity, they apparently also shared convictions of having attained that state. I have argued elsewhere that Milton in the autobiographical digressions punctuating his works rarely if ever

[46] John Conduitt, "Notes on Newton's Character," Keynes Ms. 130.07, 9r (my emphasis); https://www.newtonproject.ox.ac.uk/view/texts/normalized/THEM00169, accessed May 10, 2022.

[47] Voltaire cites the report of Newton's physician that the natural philosopher never "had any Commerce with Women" (*Letters Concerning the English Nation* [London, 1733], 116); Internet Archive, https://archive.org/details/lettersconcernin00voltuoft, accessed July 16, 2025. While the physician's authority on this question might well be questioned, we have no evidence for Newton of any romantic relationship. Iliffe voices a consensus view when he writes that Newton's "choice of a life of celibacy was an essential element of his social and religious identity" (Rob Iliffe, *Priest of Nature: The Religious Worlds of Isaac Newton* [Oxford: Oxford University Press, 2017], 17).

[48] "Fragment on the history of apostasy," Yahuda Ms. 18, 2v, National Library of Israel, Jerusalem; https://www.newtonproject.ox.ac.uk/view/texts/normalized/THEM00061, accessed May 10, 2022.

[49] "Prophecies concerning Christs 2nd coming," fol. 3, accessed December 18, 2023.

shows a consciousness of sin or follows the Pauline and Augustinian script of acknowledging sinfulness as a prelude to repentance and regenerate godliness.[50] For his part Newton, excepting the nineteen-year-old's notebook list of moral lapses, nearly all of which illustrated religious zeal rather than turpitude (e.g., "Making a mousetrap" and "Squirting water" on the Lord's Day, "Not loving Thee for Thy goodness to us"), likewise left little in the way of confession. Iliffe observes that "deep introspection with regard to identifying and repenting his sins…seem[s] to have been absent from his religious life."[51]

Newton gestured at other requirements of the prophetic interpreter in his description of "the remnant, a few scattered persons which God hath chosen"; they are "such as without being led by interest, education, or humane authorities, can set themselves sincerely & earnestly to search after truth."[52] Newton's claim that God chooses those not "led by interest" echoes Milton's argument in *The Reason of Church-Government* that those to whom God grants "in more then the scantest measure to know any thing distinctly of God, and of his true worship" are trusted with wealth "of such an orient lustre as no diamond can equall, which never the lesse he has in charge to put off at any cheap rate, yea for nothing to them that will" (*CPW* 1.801–2). The returns on the investments of knowledge are spiritual, not material. While the parable of the talents calls for the return of goods with interest, that interest is precisely not that calculated by what Milton calls "the great Marchants of this world" (*CPW* 1:802).

Newton's prophets are no more led by "humane authorities" than they are by "interest." Like Milton after him, Newton looked to Scripture first, and followed what he found in Scripture whether or not it was consistent with orthodox tradition or institutional creeds. Milton, again in the letter prefacing *De Doctrina Christiana*, insists that "I neither urge nor impose anything by any authority of my own; but on the contrary I urge everyone…that on whatever points they do not feel fully satisfied, they should withhold assent until the evidence of the scriptures prevails and induces their reason to assent and belief" (*OCW* 8.1:7).

Shared attitudes toward human authority led both Newton and Milton to question those quick to charge theological opponents with heresy. While tending in earlier work to employ "heresy" to indicate misreading of biblical teachings, even here Milton distinguished between culpable heresy, or willful distortion of the Bible, and innocent error. Eventually Milton came to see "heresy," in line with its etymological lineage from the Greek *haîresis*/*αἵρεσις*, as signifying "choice," without regard to the appropriateness or inappropriateness, or orthodoxy or

[50] Fallon, *Milton's Peculiar Grace*, 21–44.

[51] Iliffe, *Priest of Nature*, 28. Newton's 1662 list of lapses can be found in the Fitzwilliam Notebook 3r–4v, https://www.newtonproject.ox.ac.uk/view/texts/diplomatic/ALCH00069, accessed September 28, 2022.

[52] Newton, "Untitled Treatise on Revelation," 1ʳ.

unorthodoxy, of the choice. In *A Treatise of Civil Power in Ecclesiastical Causes*, he argues, "heresie, by what it signifies in that language, is no word of evil note; meaning only the choise or following of any opinion good or bad in religion or any other learning: and thus not only in heathen authors, but in the New testament without censure or blame" (*CPW* 7:247). For Milton the terms *heretic* and *heresy* in a pejorative sense should be used only for those who contradict the New Testament—"after the Evangelical books had finally been put together in writing, nothing can rightly be termed heresy except what conflicts with them" (*De Doctrina Christiana*, Epistle; *OCW* 8.1:9)—or, significantly, for those who blindly follow authority rather than their own reading of Scripture. Once the word is uncoupled from blamable misreading of Scripture, the logic of *sola scriptura* and the suspicion of human traditions leads to a more complete transvaluation of the term.[53] Choice is itself a value in contrast to accepting beliefs by tradition or by what has been inherited as orthodox, a stance of "implicit faith." Milton, employing the familiar rather than the etymological meaning of "heresy," famously in *Areopagitica* labels as "a heretic in the truth" one who "beleeve[s] things only because his Pastor says so, or the Assembly so determins, without knowing their reason, though his belief be true, yet the very truth he holds, becomes his heresie" (*CPW* 2:543).

Like Milton, Newton condemns the labeling of the unorthodox or unconventional as heresy. In the opening pages of the manuscript on Revelation, Newton counsels those who would be prophets: "as thou desirest to inherit this blessing consider & search into these Scriptures which God hath given to be a guide in these latter times, & be not discouraged by the gainsaying which these things will meet with in the world. They will call thee it may be a a Bigot, a Fanatique, a Heretique &c."[54] In his manuscript "Of the Church," Newton charged those who branded others with the label of heretic with acting contrary to the teaching of Christ: "To impose any new condition of communion wch all Christians cannot admit is to deny communion to those who by the institution of Christ have a right to it. It is to condemn those as hereticks ^whom Christ hath not condemned^ & wth what judgment ye judge ye shall be judged."[55] For both Newton and Milton, the term "heretic" has become a weapon by which the orthodox many stigmatize the prophetic few. As Stephen Snobelen has written, for Newton "orthodoxy is heresy and heresy truth."[56]

[53] For an incisive analysis of Milton's developing sense of the meaning of *heresy*, see Janel Mueller, "Milton on Heresy," in *Milton and Heresy*, ed. Stephen B. Dobranski and John Rumrich (Cambridge: Cambridge University Press, 1998), 21–38. See also David Loewenstein, *Treacherous Faith: The Specter of Heresy in Early Modern Literature and Culture* (Oxford: Oxford University Press, 2013), 282–95.

[54] Newton, "Untitled Treatise on Revelation," 4^r^.

[55] Newton, "Of the Church" (Geneva, Switzerland: Fondation Martin Bodmer), Chapter 2, fol. 42^r^.

[56] Stephen D. Snobelen, "Isaac Newton, Heretic: The Strategies of a Nicodemite," *British Journal for the History of Science* 32, no. 4 (1999): 390.

Milton and Newton, thus, shared a sense of the requirements for a prophetic interpreter of Scripture: one must strive for purity of character, especially in sexual matters, one must sufficient knowledge of languages to construe the original senses of Scriptural passages, one must labor diligently and amass learning, one must be disinterested and selfless rather than focused on gain, one must not allow tradition or human authority to color one's reading of Scripture, one must collate verses of Scripture, and one must submit one's interpretations to the test of reason. This last criterion, as I will suggest in Chapter 4, is one they shared with Socinians.

Prophetic Practice

When Milton and Newton, both independent-minded, strong rationalists raised on Puritan principles at home and at Cambridge, read the Bible, they found similar things. Some are not surprising, as they are the common currency of Reformed thinking. But they also found in Scripture no support for virtually universal Christian beliefs, most notably the trinitarian belief in the co-equality and co-eternity of the Son of God. They also shared an opposition to the common belief in the natural immortality of the soul. They identified the same archaeology of these errors, in the infiltration of primitive Christian theology by Greek philosophical models and terms.[57] Their shared skepticism concerning several common Christian dogmas is the subject of the next chapter.

While Newton is adamant on the unique divinity of the Father and on the inferiority of the Son and Holy Spirit, in regard to most other doctrinal points he is, unlike Milton, either agnostic or non-dogmatic. In addition to shared antitrinitarianism and mortalism, several of Milton's prophetic interpretations fall under what Newton would regard as *adiaphora*, as things indifferent. In terms drawn from 1 Corinthians 3:1–2, Newton distinguished the *adiaphora* as "strong meat" for men of more mature minds, who might disagree on matters not necessary for salvation, from the "milk for babes," those things to be taught and accepted before baptism:

> by faith we are to understand a belief not of all the truths conteined in scripture, but only of certain fundamental points relating to practice. For there were certain principles of religion preached by the Apostles from the beginning, & in which all men were to be instructed before baptism by catechizing them, & there were many other truths which they were to learn after baptism by studying the scriptures to their lives end of their lives. The Principles the Apostle compares to

[57] See Paul C. H. Lim, *Mystery Unveiled: The Crisis of the Trinity in Early Modern England* (Oxford: Oxford University Press, 2012), 28, 36.

> milk to be given to babes, the other truths to strong meat wch belongeth to men of full age.... & they [the milk for babes] are repentance from dead works, faith in God, Baptism, laying on of hands, & the doctrine of the resurrection & day of judgment upon wch repentance from dead works is grounded.[58]

Significantly, the "strong meats" that Newton, however personally combative, brackets as not necessary to salvation, contain many doctrinal points of absolutely central importance to Milton. In addition to the "milk for babes," Newton writes,

> there are \in the scriptures/ many truths of great importance but more difficult to be understood & not \so/ absolutely necessary to salvation.... And to these truths I refer *what Christ did before his incarnation* and between his death & resurrection, what he doth now in heaven & \how the saints shall reign with him as [*illeg*] Kings and Priests in the day of judgment & rule the nations wth a rod of iron &/what he or they shall do after the day of judgment, as also *all disputable questions about Providence, Predestination, free Will, Grace, the origin of evil,* the \nature of the/ satisfaction made by Christ, *the nature of Angels, the state of the dead between death & the resurrection.*[59]

The italicized doctrinal articles here are all of real interest to Milton and some, especially those surrounding providence, predestination, and free will, he does not see as adiaphora but instead considers of central importance to faith (if not necessarily required for salvation). We will see that in their interpretation of identical biblical verses Milton will focus on—and Newton ignore—some of these doctrinal points.

The greatest difference between Milton and Newton as prophets is in their understanding of the relation between prophecy and history. Newton had for decades cultivated an intense interest in universal history. For Newton history was central to his prophetic practice, as is starkly evident in two books published shortly after his death in 1727, *The Chronology of Ancient Kingdoms Amended* (1728) and *Observations upon the Prophecies of Daniel and the Apocalypse of St. John* (1733).[60] In these books and in his prophetic manuscripts, Newton labored to demonstrate both the authority of Scriptural histories in relation to pagan histories and the accuracy of biblical prophecies of subsequent history. Milton shared an interest in history, having published a *History of Britain* in 1670

[58] Newton, "Of the Church," Chapter 2, fol. 20^{r}.

[59] Newton, "Of the Church," Chapter 3, fols. 23^{r}–24^{r}, my emphasis.

[60] The late seventeenth- and early eighteenth-century manuscripts and notes for the *Chronology* can be found in the National Library of Israel's Yahuda manuscripts and in the libraries of Cambridge University and New College, Oxford. Yahuda Ms. 25.1a I, "Draft sections of the 'Chronology of Ancient Kingdoms Amended'"; Yahuda Ms. 25.2d, "Drafts on chronology"; New College Library Ms. 361(3), "Papers relating to chronology and 'Theologiæ Gentilis Origines Philosophicæ'"; and Cambridge University Library Ms Add. 3988, "Fair copies of the 'Short Chronicle' and 'Chronology of Ancient Kingdoms Amended.'" All can be found in The Newton Project.

and composed *A Brief History of Moscovia*, published posthumously in 1682, but he did not share Newton's commitment to rationalizing universal history and demonstrating its prediction in the prophetic books of Scripture.

In his various drafts and in the published *Chronology of Ancient Kingdoms Amended* Newton compiles tables of comparative chronology to demonstrate the compatibility of the ancient chronologies of Gentile nations, once corrected and adjusted, with biblical chronology. This involved foreshortening what he saw as exaggerated claims for the antiquity of cultures other than the Hebrew culture. As Newton notes, "Greek Chronologers" typically overestimated the lengths of monarch's reigns as running thirty-five to forty years, substituting as more reasonable reigns of eighteen to twenty years.[61] In addition, Newton applied astronomical calculations to correct ancient chronologies; he calculated, for example, the precession of the equinoxes to redate the journey of the Argonauts to a significantly later date, after the death of Solomon, than that claimed in Greek chronologies.[62] While Newton gave precedence to the historical books of the Hebrew Scriptures among ancient historical texts, he was unusual in his understanding of these books as works of history composed by men rather than directly inspired by God. According to Manuel, "Newton came to distinguish rather sharply between two types of books in the Biblical canon; those that were narrative-historical and those that were direct prophecy, the word of the living God. By his middle years Newton had come to believe that Biblical descriptions of historical events were written for the most part by contemporaries of those events."[63] Newton believed that Moses, the one canonical writer not contemporary with all the events of the books he composed, had access to the most ancient records. The greater precision in dating in the Hebrew scriptures, in Newton's view, came from the contemporaneity of the writing and the fact of their careful preservation, owing to their continued use in Jewish liturgy, of the texts of canonical works. Of the truly prophetic works, the Books of Daniel and Revelation were of the greatest interest to Newton, who saw them as closely linked in purpose and content. They are set apart not only by direct divine inspiration but by their highly allegorical language and their shared prophetic vocabulary.

In his *Observations*, Newton makes clear the connection between his interests in biblical prophecy and in history, as well as the prophetic character of his role as exegete and historian. The Books of Daniel and Revelation are intimately linked: "This Prophecy is called *the Revelation*, with respect to *the scripture of truth*, which *Daniel* was commanded to *shut up and seal, till the time of the end. Daniel* sealed it *until the time of the end*; and until that time comes, the Lamb is opening

[61] Newton, *The Chronology of Ancient Kingdoms Amended* (London, 1728), 52. See Richard Westfall, *Never at Rest: A Biography of Isaac Newton* (Cambridge: Cambridge University Press, 1980), 813.

[62] Westfall, *Never at Rest*, 813.

[63] Manuel, *Religion of Isaac Newton*, 84.

the seals."[64] The prophecies of Daniel, which Spinoza saw as muddled because of the mental limitations of the prophet, are for Newton obscure by design: "*In the time of the end the wise shall understand, but none of the wicked shall understand.*"[65] Daniel prophesied the future until the end of time, as did John in Revelation later. The coming of Christ verified much of Daniel's prophecy, and both Daniel and Revelation would be verified by the Second Coming. As that Second Coming approached, the prophetic few would understand and teach the truths hidden in the prophetic books:

> these Prophecies of *Daniel* and *John* should not be understood till the time of the end: but then some should prophesy out of them in an afflicted and mournful state for a long time, and that but darkly, so as to convert but few. But in the very end, the Prophecy should be so far interpreted as to convince many. *Then*, saith *Daniel, many shall run to and fro, and knowledge shall be encreased* [Dan 12:4].[66]

Newton believed that he was living in the times approaching "the last age," when all would be made clear by those with prophetic insight into the prophecies. He makes this point most clearly in his manuscript that he labeled the "Irenicum":

> And as the prophesies of the old Testament remained in obscurity till Christs first coming & then were interpreted by Christ & the interpretations became the religion of the Christians: so the prophesies of both Testaments relating to Christs second coming may remain in obscurity till that coming, & then be interpreted by divine authority, & the interpretations become the religion of God's people till Christ has put all things under his feet in heaven & earth & shall deliver up the kingdom to the father.[67]

The prophetic line here is striking. Newton, as will appear in a moment, thought of Christ both as a particular focus of worship and as one more prophet in a long line of prophets who arose when nations fell away from the true worship of the primitive monotheistic religion of the one omnipotent God. In this passage from the "Irenicum," the coming of Christ verifies the prophecies of the Old Testament, including Daniel, and the second coming will verify the prophecies of the New. In the meantime, as the Second Coming approaches the scriptural prophecies will be "interpreted by divine authority, & the interpretations [will] become the

[64] Newton, *Observations upon the Prophecies*, 247–48 (II.1).
[65] Newton, *Observations upon the Prophecies*, 249 (II.1).
[66] Newton, *Observations upon the Prophecies*, 248 (II.1).
[67] Newton, "Irenicum, or Ecclesiastical Polyty tending to Peace," Keynes Ms. 3, King's College, Cambridge, 14r, my emphasis; https://www.newtonproject.ox.ac.uk/view/texts/normalized/THEM00003, accessed January 23, 2024.

religion of God's people" until the end of time. In other words, there will be latter-day prophets, among whom Newton, by his practice, includes himself. On the one hand, Newton interprets obscure passages, the meaning of which by design remains obscure to "the wicked," as one who prepares the way for the coming of Christ; on the other hand, he plays a role analogous to Christ's, in verifying earlier prophecies. And the ultimate motivation for this verification is the demonstration of the absolute omnipotence of God, who governs human history so as to fulfill the events that he prophesies through the writings of the prophet Daniel and John the Evangelist.

If Newton places himself in a Christ-like prophetic role, he also expands the line of prophets outside the Judaeo-Christian canon and tradition. In his manuscript "A short Schem of the true Religion," Newton includes Christ in a catalogue prophets who articulated and/or revived these simple precepts of the true religion: "We must love our neighbour as our selves, we must be charitable to all men for charity is the greatest of graces, greater then even faith or hope & covers a multitude of sins. We must be righteous & do to all men as we would they should do to us."[68] This catalogue, significantly, contains not only biblical figures but even Socrates and Confucius: "This was the ethics morality or good manners taught the first ages by Noah & his sons in some of their seven precepts the later heathens by Socrates Cicero Confutius & other Philosophers, the Israelites by Moses & the Prophets & the Christians more fully by Christ & his Apostles. This is that law which the Apostle tells you was written in the hearts of the Gentiles & *by which they are to be judged in the last day* Rom. 2.12, 14, 15."[69]

The bulk of Newton's work on Daniel and Revelation is the meticulous historical work retrospectively demonstrating the point for point accuracy of the obscure historical prophecies of the scriptural prophets. For Newton, the key to reading prophecy is to look back, not forward:

> The folly of Interpreters has been, to foretel times and things by this Prophecy, as if God designed to make them Prophets. By this rashness they have not only exposed themselves, but brought the Prophecy also into contempt. The design of God was much otherwise. He gave this and the Prophecies of the Old Testament, not to gratify men's curiosities by enabling them to foreknow things, but that after they were fulfilled they might be interpreted by the event, and his own Providence, not the Interpreters, be then manifested thereby to the world. For the event of things predicted many ages before, will then be a convincing argument that the world is governed by providence.[70]

[68] Newton, "A Short Schem of the True Religion," 1^r–2^r; https://www.newtonproject.ox.ac.uk/view/texts/normalized/THEM00007; accessed December 18, 2023.

[69] Newton, "Short Schem," 2^r; my emphasis.

[70] Newton, *Observations upon the Prophecies*, 249-50 (II.1).

History's congruence with the prophecies of Daniel and John demonstrates not only God's providence but also God's omnipotence. All will happen, and all has happened, as an all-powerful God has ordained. It is presumptuous to try to predict the future, and God's prophecies are purposely obscure to discourage bold speculation. This did not prevent many of Newton's contemporaries from calculating the time of the Second Coming, but Newton's few ventures into this area were surrounded by qualifications, and a tendency to push the date back rather than bring it forward.[71]

The marked difference between Milton's and Newton's reading of the prophetic works can be gauged by their strikingly different uses of the same chapters and verses. Newton goes to extraordinary lengths, for example, to coordinate the prophecy in Daniel 7:7–9 of the ten-horned beast with subsequent history:

> 7 After this I saw in the night visions, and behold a fourth beast, dreadful and terrible, and strong exceedingly; and it had great iron teeth: it devoured and brake in pieces, and stamped the residue with the feet of it: and it *was* diverse from all the beasts that *were* before it; and it had ten horns. 8 I considered the horns, and, behold, there came up among them another little horn, before whom there were three of the first horns plucked up by the roots: and, behold, in this horn *were* eyes like the eyes of man, and a mouth speaking great things. 9 I beheld till the thrones were cast down, and the Ancient of days did sit, whose garment *was* white as snow, and the hair of his head like the pure wool: his throne *was like* the fiery flame, *and* his wheels *as* burning fire.

Chapter 6 of the *Observations*, "Of the ten Kingdoms represented by the ten horns of the fourth Beast," explicates the last five words of Daniel 7:7 ("and it had ten horns") over 27 pages, laying out in exhaustive detail the kingdoms represented by each horn:

> Now by the wars above described the *Western* Empire of the *Romans*, about the time that Rome was besieged and taken by the *Goths*, became broken into the following ten kingdoms.
>
> 1. The kingdom of the *Vandals* and *Alans* in *Spain* and *Africa*.
> 2. The kingdom of the *Suevians* in *Spain*.
> 3. The kingdom of the *Visigoths*.
> 4. The kingdom of the *Alans* in *Gallia*.
> 5. The kingdom of the *Burgundians*.
> 6. The kingdom of the *Franks*.
> 7. The kingdom of the *Britains*.

[71] See Stephen Snobelen, "'A Time and Times and the Dividing of Time' Isaac Newton, the Apocalypse, and 2060 A. D.," *Canadian Journal of History* 38 (2003): 550–51.

8. The kingdom of the *Hunns.*
9. The kingdom of the *Lombards.*
10. The kingdom of *Ravenna.*[72]

To convey the level of attention to historical detail, a long quotation of one of the ten kingdoms, though the shortest of the group, may be necessary:

1. The Kings of the *Vandals* were, A.C. 407 *Godegesilus*, 407 *Gunderic*, 426 *Geiseric*, 477 *Hunneric*, 484 *Gundemund*, 496 *Thrasamund*, 513 *Geiseric*, 530 *Gelimer*. *Godegesilus* led them into *Gallia* A.C. 406, *Gunderic* into *Spain* A.C. 409, *Geiseric* into *Africa* A.C. 427; and *Gelimer* was conquered by *Belisarius* A.C. 533. Their kingdom lasted in *Gallia*, *Spain* and *Africa* together 126 years; and in *Africa* they were very potent. The *Alans* had only two Kings of their own in *Spain*, *Resplendial*, and *Ataces*, *Utacus* or *Othacar*. Under *Resplendial* they went into *France* A.C. 407, and into *Spain* A.C. 409. *Ataces* was slain with almost all his army by *Vallia* King of the *Visigoths* A.C. 419. And then the remainder of these *Alans* subjected themselves to *Gunderic* King of the *Vandals* in *Boetica*, and went afterwards with them into *Africa*, as I learn out of *Procopius*. Whence the Kings of the *Vandals* styled themselves Kings of the *Vandals* and *Alans*; as may be seen in the Edict of *Hunneric* recited by *Victor* in his *Vandalic* persecution. In conjunction with the *Chatti*, these *Alans* gave the name of *Cathalaunia*, or *Catth-Alania*, to the Province which is still so called. These *Alans* had also *Gepides* among them; and therefore the *Gepides* came into *Pannonia* before the *Alans* left it. There they became subject to the *Hunns* till the death of *Attila* A.C. 454, and at length were conquered by the *Ostrogoths*.[73]

Newton then goes on in the next, sixteen-page chapter, to interpret the little, eleventh horn in Daniel 7:8, which turns out to be the Church of Rome. The "eyes like the eyes of man, and a mouth speaking great things" point to the Pope: "A Seer, **ἐπίσκοπος** is a Bishop in the literal sense of the word; and this Church claims the universal Bishoprick. ¶ With his mouth he gives laws to kings and nations as an Oracle; and pretends to Infallibility, and that his dictates are binding to the whole."[74]

It is not merely that Milton is less meticulous in unraveling the historical prophecies in Daniel. He has no interest in the subject at all. In citing Daniel 7, he ignores verses 7 and 8 entirely, picking up instead at verses 9 and 10, which he cites as illustrating "that supreme excellence of God by virtue of which he is truly

[72] Newton, *Observations upon the Prophecies*, 47 (I.6).
[73] Newton, *Observations upon the Prophecies*, 48–49 (I.6).
[74] Newton, *Observations upon the Prophecies*, 75 (I.7). Milton earlier had associated prelates with eyes and mouths in "Lycidas"; "Blind mouths! that scarce themselves know how to hold / A sheep-hook, or have learned aught else the least / That to the faithful herdman's art belongs!" (119–21).

perfect, and in supreme glory truly blessed, and by supreme justness and his own merit sovereign Lord of all" (*DDC* I.2; *OCW* 8.1:47). What interests Newton in Daniel 8 is the vision of the ram and the goat, which for Newton are the kingdoms respectively of the Medes and the Persians and of the Greeks.[75] Milton again ignores the verses on the ram and the he-goat. He does cite Daniel 8:13 ("Then I heard one saint speaking, and another saint said unto that certain *saint* which spake, How long *shall be* the vision *concerning* the daily *sacrifice*, and the transgression of desolation, to give both the sanctuary and the host to be trodden under foot?"), which he interprets as evidence of the extent and nature of angels' knowledge: "The good angels do not see into all things in respect of God, as the Papists imagine, but [see] only, through revelation, those things which God has decided to [reveal], and other things through a certain understanding process of reasoning; but many things they do not know, for an inquisitive angel asks, Dan. 8: 13: *for how long will this vision last?*" (*DDC* I.9; *OCW* 8.1.351). Newton, on the other hand and characteristically, mines the verse for historical confirmation of God's power:

> *Then I heard*, saith [8] *Daniel, one saint speaking, and another saint said unto that certain saint which spake, How long shall be the vision concerning the daily sacrifice, and the transgression of desolation, to give both the sanctuary and the host to be trodden under foot? And he said unto me, Unto two thousand and three hundred days; then shall the sanctuary be cleansed.* Daniel's days are years; and these years may perhaps be reckoned either from the destruction of the Temple by the *Romans* in the reign of *Vespasian*, or from the pollution of the Sanctuary by the worship of *Jupiter Olympius*, or from the desolation of *Judea* made in the end of the *Jewish* war by the banishment of all the *Jews* out of their own country, or from some other period which time will discover. Henceforward the last horn of the Goat continued mighty under the *Romans*, till the reign of *Constantine* the great and his sons: and then by the division of the *Roman* Empire between the *Greek* and *Latin* Emperors, it separated from the *Latins*, and became the *Greek* Empire alone, but yet under the dominion of a *Roman* family; and at present it is mighty under the dominion of the *Turks*.[76]

While Milton is keenly interested in typological relations between the Hebrew Scriptures and the Greek Testament, and in drawing on both for models for faith and action, he shows no interest in plotting the sweep of human history as evidence of the veracity of divine prophecy. He comes closest when he employs Daniel 9:24 ("Seventy weeks are determined upon thy people and upon thy holy city, to finish the transgression, and to make an end of sins, and to make reconciliation

[75] Newton, *Observations upon the Prophecies*, 115 (I.9).
[76] Newton, *Observations upon the Prophecies*, 122 (I.9).

for iniquity, and to bring in everlasting righteousness, and to seal up the vision and prophecy, and to anoint the most Holy") to demonstrate the supercesssion of Judaism by Christianity with the coming of the Messiah: "That the Messiah has already come is proved, against Jewish belief,...it was being predicted that his advent would occur while the second temple and the Jewish state were [still] standing" (*DDC* I.14; *OCW* 8.1:487–89). Milton's interest is in the doctrinal priority of the Christian faith rather than demonstration of God's power in foreseeing (and directing) human history. Newton devotes to the same verse 560 words of dense historical detail.[77]

The pattern repeats itself in Milton's and Newton's use of Revelation. To take one example, Newton interprets Revelation 5:8 as an historical description of "primitive Christians": "*And when he had taken the book, the four Beasts and four and twenty Elders fell down before the Lamb, having every one of them harps, and golden vials full of odours, which are the prayers of saints*.... The Beasts and Elders therefore represent the primitive *Christians* of all nations; and the worship of these *Christians* in their Churches is here represented under the form of worshiping God and the Lamb in the Temple."[78] Milton cites the same verse, without feeling the need to tie it to any historical period or event; instead, he cites it as evidence of the efficacy of petitionary prayer through Christ: "Petition is that whereby, at the prompting of the holy spirit [and] on the basis of faith in Christ, we reverently request from God any lawful thing whatever for ourselves and for others....Rev. 5: 8" (*DDC* II.4; *OCW* 8.2:969–71).

The two differ similarly in their reading of the beginning of Revelation 9:

> 1 And the fifth angel sounded, and I saw a star fall from heaven unto the earth: and to him was given the key of the bottomless pit. 2 And he opened the bottomless pit; and there arose a smoke out of the pit, as the smoke of a great furnace; and the sun and the air were darkened by reason of the smoke of the pit. 3 And there came out of the smoke locusts upon the earth: and unto them was given power, as the scorpions of the earth have power.

Milton cites Revelation 9:1 only as an indication of one of the names of Hell or the "place of punishment" (*DDC* I.33; *OCW* 8.2:893). Newton, on the other hand, unpacks the passage for its historical signification. Again, only a long (but still partial) quotation can communicate the meticulous, even obsessive effort to relate prophecy to history:

> The fifth trumpet sounded to the wars, which the *King of the* South, as he is called by *Daniel*, made *in the time of the end*, in *pushing at the King who did according to his will*. This plague began with the *opening of the bottomless pit*,

[77] Newton, *Observations upon the Prophecies*, 129–32 (I.10).
[78] Newton, *Observations upon the Prophecies*, 260 (II.2).

which denotes the letting out of a false religion: the *smoke which came out of the pit*, signifying the multitude which embraced that religion; and the *locusts which came out of the smoke*, the armies which came out of that multitude. This pit was opened, to let out smoke and locusts into the regions of the four monarchies, or some of them. *The King of these locusts* was the *Angel of the bottomless pit*, being chief governor as well in religious as civil affairs, such as was the Caliph of the *Saracens*. Swarms of locusts often arise in *Arabia fœlix*, and from thence infest the neighbouring nations: and so are a very fit type of the numerous armies of *Arabians* invading the *Romans*. They began to invade them A.C. 634, and to reign at *Damascus* A.C. 637. They built *Bagdad* A.C. 766, and reigned over *Persia*, *Syria*, *Arabia*, *Egypt*, *Africa* and *Spain*. They afterwards lost *Africa* to *Mahades*, A.C. 910; *Media*, *Hircania*, *Chorasan*, and all *Persia*, to the *Dailamites*, between the years 927 and 935; *Mesopotamia* and *Miafarekin* to *Nasiruddaulas*, A.C. 930; *Syria* and *Egypt* to *Achsjid*, A.C. 935, and now being in great distress, the Caliph of *Bagdad*, A.C. 936, surrendred all the rest of his temporal power to *Mahomet* the son of *Rajici*, King of *Wasit* in *Chaldea*, and made him Emperor of Emperors. But *Mahomet* within two years lost *Bagdad* to the *Turks*; and thenceforward *Bagdad* was sometimes in the hands of the *Turks*, and sometimes in the hands of the *Saracens*, till *Togrul-beig*, called also *Togra*, *Dogrissa*, *Tangrolipix*, and *Sadoc*, conquered *Chorasan* and *Persia*; and A.C. 1055, added *Bagdad* to his Empire, making it the seat thereof. His successors *Olub-Arslan* and *Melechschah*, conquered the regions upon *Euphrates*; and these conquests, after the death of *Melechschah*, brake into the kingdoms of *Armenia*, *Mesopotamia*, *Syria*, and *Cappadocia*. The whole time that the Caliphs of the *Saracens* reigned with a temporal dominion at *Damascus* and *Bagdad* together, was 300 years, viz. from the year 637 to the year 936 inclusive. Now locusts live but five months; and therefore, for the decorum of the type, these locusts are said to *hurt men five months and five months*, as if they had lived about five months at *Damascus*, and again about five months at *Bagdad*; in all ten months, or 300 prophetic days, which are years.[79]

For Newton the key lesson of the opening of the ninth chapter of Revelation is not an indication of the final fate of the sinful but a precisely though obscurely accurate prophecy of the rise of Islam and the strife between rival infidels.

Milton's focus in his prophetic interpretation of Scripture is access to true doctrine and especially to the moral life of the believer and the exercise of free will. His focus, that is, is on the individual, where Newton's is on peoples, monarchies, and the large historical movements that confirm divine prophecy and thus divine omnipotence. There is perhaps no better example of the radically different prophetic agendas than their responses to Revelation 2:10: "Fear none

[79] Newton, *Observations upon the Prophecies*, 301–3 (II.3).

of those things which thou shalt suffer: behold, the devil shall cast some of you into prison, that ye may be tried; and ye shall have tribulation ten days: be thou faithful unto death, and I will give thee a crown of life." For Newton, this verse is an historically specific and precise prophecy of the persecution of Christians by Diocletian in 302–12 CE.

> The persecution of *Dioclesian* began in the year of *Christ* 302, and lasted ten years in the *Eastern* Empire and two years in the *Western*. [Here Newton quotes Rev. 2:9–10). The tribulation of ten days can agree to no other persecution than that of *Dioclesian*, it being the only persecution which lasted ten years. By *the blasphemy of them which say they are* Jews *and are not, but are the synagogue of* Satan, I understand the Idolatry of the *Nicolaitans*, who falsly said they were *Christians*.[80]

Milton, on the other hand, finds in the verse confirmation of his beliefs in free will and individual responsibility. John's prediction that "the devil shall cast some of you into prison" is for Milton a prophecy that Christians will face "Good temptation" (*Bona tentatio*),

> whereby God tempts even good people for the sake of proving them, not for his own sake—as if he did not know what they would be like—but in order either to train or else highlight their faith, or else their patience, as he tempted Abraham and Job, or in order to lessen their assurance and show up their weakness, so that both they themselves may know themselves more truly, and others may be instructed. (*DDC* I.8; *OCW* 8.1:335)

The contrast with Newton here is particularly strong. While Newton numbered predestination among the strong meats belief in which is not necessary for salvation, his emphasis on God's unlimited omnipotence moved him toward acceptance of a Calvinist view of predestination. He emphasizes God's power and God's mercy on those chosen for salvation, with no hint of Milton's grappling with the problem for theodicy in God's predestinating some to damnation. Newton instead focuses on what God must endure from the vessels of wrath: "What if God willing to shew his wrath & to make his power known, endured with much long suffering the vessels of wrath filled to destruction? And that he might make known the riches of his glory on the vessels of mercy which he had afore prepared unto glory; even us whom he hath called not of the Iews onely but

[80] Newton, *Observations upon the Prophecies*, 287–88 (II.3).

also of the Gentiles. Rom 9.11."[81] Milton on the other hand wrestles mightily with a passage from Romans that suggests particular predestination:

> you will say: *it is not a matter of one who wishes, nor of one who runs, but of God who is merciful*, Rom. 9: 16. Yet, I reply, we introduce neither one who wishes nor one who runs, but one who is less unwilling, less hesitant, less resistant, and God who is no less merciful, and at the same time supremely wise and supremely just. Meanwhile a person who says it is not a matter of one who wishes, nor of one who runs, surely assumes one who wishes, and assumes one who runs, but achieves just this, that he grants neither to the one who wishes nor to the one who runs any praise or merit. But when God had decided to restore humankind, he also indubitably decreed (for what was fairer?) to restore lost freedom to the will, in some measure at least. (*DDC* I.4; *OCW* 8.1:95)

Milton carefully separates his position from Pelagianism; one cannot will or run without God's grace. But he proposes that a just God recognizes those who are "less unwilling" and gives them the assistance necessary to will and run. Otherwise, he argues, we must imagine a God who makes no distinction between those who by the exercise of free will are either firmly unwilling or "less unwilling." For Newton, God knows what individuals will choose because those choices, like everything else, are determined by God. Milton's God tests free individuals; if he knows how they will choose, those choices are logically prior to God's knowledge of them; God foresees free choices: "Nor in truth has anything happened because God foresaw it, but rather God has foreseen each event, since it will turn out thus because of its own causes through his decree for free agents, causes well known to himself; and the outcome does not in fact rest with God who foresees it but solely with the agent who is foreseen" (*DDC* I.3; *OCW* 8.1:63).

The end of Revelation 2:10, "be thou faithful unto death, and I will give thee a crown of life," Milton relates to his anti-Calvinist, Arminian insistence that there is no predestined election to perseverance: "For this reason, not *the chosen* but *those who persevere* are said to attain salvation" (*DDC* I.25; *OCW* 8.2:665). For Milton all rational creatures, angels as well as human beings, are free to choose evil or against God until the end of time. As Raphael tells Adam in *Paradise Lost*,

> Myself and all th' angelic host that stand
> In sight of God enthroned, our happy state
> Hold, as you yours, while our obedience holds;

[81] Newton, "Theological Notebook, Part One," Keynes Ms. 2, 31^{r}; https://www.newtonproject.ox.ac.uk/view/texts/normalized/THEM00180, accessed March 18, 2024.

On other surety none; freely we serve,
Because we freely love, as in our will
To love or not; in this we stand or fall. (*Paradise Lost* 5.535–40)

Milton's prophetic role, his self-imposed imperative to be "doctrinal and exemplary to a nation" (*Reason of Church-Government*; *CPW* 1:815), takes the form of interpreting Scripture as a guide to the choices of free individuals, whether the choice is what to believe, how to discern good amid the myriad appearances of evil, or how to worship God. While Newton's personal scrupulousness is a match for Milton's, Newton's prophetic focus, on the other hand, is on large scale movements of peoples and kingdoms, the shape of which is overseen by an omnipotent God just as are the movements of the heavens.

While their diverging interests account for their markedly different interpretation of Daniel and Revelation, Milton and Newton converged in regarding themselves as members of prophetic few who exhorted fellow Christians to return to the primitive purity of the Church. While Moses exclaims "would God that all the LORD'S people were prophets, *and* that the LORD would put his spirit upon them" (Numbers 11:29), Milton and Newton inclined toward the opening of Isaiah ("Except the Lord of hosts had left unto us a very small remnant, we should have been as Sodom," Is 1:9). They agreed also on many of illegitimate accretions to primitive and pure Christian belief. In the next chapter, I will explore their shared Arianism and their affinities with Socinianism. Even as they agree on several Christological principles, one can see the handprints of their differing prophetic methods and their not merely different but opposing theological imperatives.

Milton and Newton: Matter, God, and Modernity. Stephen M. Fallon, Oxford University Press.
 DOI: 10.1093/9780198988939.003.0004

4
From Arianism to Socinianism

Late in 1675, a year after Milton's death, Daniel Skinner, having obtained the manuscript of *De Doctrina Christiana*, sent it for publication to Daniel Elzevier in Amsterdam. Fifteen years later, on November 14, 1690, Isaac Newton sent to his friend John Locke *An historical account of two notable corruptions of Scripture, in a Letter to a Friend,* with the understanding that Locke would forward the text to Jean LeClerc in Amsterdam for publication. The texts share more than a route from England to Amsterdam. Both espouse an antitrinitarian and Arian understanding of the Son of God as neither co-eternal nor co-essential with the Father, and neither was published as intended by the sender—Milton's because the theologian Phillipp van Limborch advised Elzevier against publishing a work in which "the strongest Arianism was to be found throughout,"[1] and Newton's because the temperamentally cautious and belatedly alarmed author instructed Locke to stop the publication.[2] The manuscripts languished for long periods before being rediscovered and published. Newton's was found in Amsterdam's Remonstrants Library and published in 1754. Milton's would wait another seventy years before it was discovered in the Old State Papers Office in Whitehall and published in 1825. The texts, potentially dangerous to their authors, make clear that these two prophetic interpreters of the Bible rejected a key teaching of orthodox Christianity.

In this chapter, I will first compare the demonstrably Arian beliefs of Milton and Newton before exploring how they in overlapping ways approached Socinian positions. I will chart the movements of Milton and Newton at the borders of Arian and Socinian understandings of the Son of God and examine the implications of those movements. The border was permeable in the seventeenth century, and observers sometimes confused the Arian and Socinian positions, because Arians and Socinians shared salient elements of heretical antitrinitarian Christology.[3] Both Arians and Socinians denied the co-eternity, co-essentiality, and co-equality of the Father and the Son, central tenets of orthodox trinitarianism. The unity of the three persons of Father, Son, and Spirit is viewed by Arians and Socinians as a

[1] Quoted in Gordon Campbell, Thomas N. Corns, John K. Hale, and Fiona J. Tweedie, *Milton and the Manuscript of "De Doctrina Christiana"* (Oxford: Oxford University Press, 2010), 7.

[2] Richard Westfall, *Never at Rest: A Biography of Isaac Newton* (Cambridge: Cambridge University Press, 1980), 490–91.

[3] For a view of Milton's and Newton's relationship to Arianism and Socinianism different from this chapter's, see John Rogers, "Newton's Arian Epistemology and the Cosmogony of *Paradise Lost*," *ELH* 86, no. 1 (2019): 77–106.

unity or concurrence of wills rather than a metaphysical union of three persons in one God, a mystery embraced by trinitarians.

Some housekeeping is necessary at the beginning. The terms *Arian* and *Socinian* must be handled differently as labels. We know the writings of the fourth-century Arius primarily through their fragmentary quotation in the works of two contemporary and several later theologians hostile to his denial of the Trinity, and in particular his bitter enemy Athanasius, whose trinitarian view of the Godhead prevailed at Council of Nicea in 325 CE.[4] Given the brief and fragmentary written remains of Arius, available to us only in the writings of hostile witnesses, we cannot with any confidence reconstruct a systematic Arian theology. We can, nevertheless, identify the key Christological idea associated with Arius that Milton, Newton, and their contemporaries knew by the name of Arianism.

In the "Thalia," a verse popularization of Arius' antitrinitarianism that may or may not have been written by Arius, one can find the kernel of what Arianism meant to Milton and his contemporaries:

> And so God Himself, as he really is, is inexpressible to all.
> He alone has no equal, no one similar (*homoios/ὅμοιος*), and no one of the same glory.
> We call him unbegotten, in contrast to him who by nature is begotten.
> We praise him as without beginning in contrast to him who has a beginning.
> We worship him as timeless, in contrast to him who in time has come to exist.
>
> He who is without beginning made the Son a beginning of created things
> He produced him as a son for himself by begetting him.
> He [the son] has none of the distinct characteristics of God's own being (*hypostasis/ὑπόσταοις*)
> For he is not equal to, nor is he of the same being (*homoousios/ὁμοούσιος*) as him.
> ....
> As far as their glories, one infinitely more glorious than the other.
> The Father in his essence (*ousia/οὐσία*) is a foreigner
> (xenos/*ξένος*) to the Son, because he exists without beginning.

[4] The scant literary remains of Arius consist of three surviving letters and a versification of doctrine attributed to Arius: his letter to Alexander of Alexandria, quoted in Athanasius' *On the Councils of Arminum and Seleucia*, 16; his letter to Eusebius of Nicomedia, recorded in Epiphanius' *Refutation of All Heresies*, 69.6 and Theodoret's *Church History* i.5; his confession to Constantine, recorded in Socrates Scholasticus' *Church History* 1.26.2 and Sozomon's *Church History* 2.27.6–10; and the *Thalia* (see R. P. C. Hanson, *The Search for the Christian Doctrine of God: The Arian Controversy 318–381* [London: T & T Clark, 2005], 5–18).

> Understand that the Monad [eternally] was; but the Dyad was
> not before it came into existence.
> It immediately follows that, although the Son did not exist, the
> Father was still God.[5]

In these lines, more than a third of the whole, one finds the key elements belonging to Arianism as understood in the seventeenth century. God the Father is the only unbegotten, eternal being; the Son, because begotten, has by definition a beginning and is thus not eternal; the Son does not share the Father's essence nor his characteristics (or perfections); and the Son, far from being one in essence with the Father, is other from the Father (the language is vivid here; the Father is a "foreigner/*xénos*/ξένος"). Explicitly denying the terms of Athanasius and the Nicene Creed, the Son is not one "of the same being (*homoousios*/ὁμοούσιος) as him" (or, that is, "one in being with the Father").

If we have little and as far as I know no writing from Arius apart from that recorded in the works of those condemning his works as heretical, the case could not be more different with Fausto Sozzini, or Socinius, the sixteenth-century Italian theologian from whose name the label Socinian derives. Sozzini left extensive theological writings, and his thought was elaborated by a series of followers, labelled Socinians, Polish Brethren, or Unitarians. Socinian theology is laid out in key Socinian texts, such as the 1605 *Katechizm Rakowski* or *Racovian Catechism*, licensed for publication in England by Milton in 1650,[6] and Johannes Crellius' 1631 *De uno Deo patre libri duo, in quibus multa etiam de filii Dei & Spiritus Sancti disseruntur*. It comprises an array of doctrinal and ecclesiological positions. While Socinians shared antitrinitarianism with Arians, they embraced an additional set of distinctive positions, including belief that the Son of God came into being not before Creation but at the Incarnation; rejection of the Aristotelian and Athanasian distinction between essence and particular person (or between *ousia*/οὐσία and *hypostasis*/ὑπόστασις) that underlay trinitarian thought; the position that God is God by virtue primarily of his dominion rather than his essence; denial of natural immortality; emphasis on rationally explicable divine justice; a conviction that primitive doctrine is preserved by a remnant; insistence on believer's baptism; and acceptance of mortalist and in some cases materialist conceptions of the soul. Skeptical about mystery, Socinians emphasized the reasonableness of scriptural truths and the perspicuousness of Scripture, and they minimized or denied the judicial atonement theory of redemption. While, as is

[5] Arius, "Thalia," trans. Aaron J. West, *Fourth Century Christianity, Wisconsin Lutheran College*; https://www.fourthcentury.com/arius-thalia-intro/, accessed August 6, 2020.

[6] The book was published by William Dugard, who, along with Milton, was interrogated by the Committee of Parliament for the Propagation of the Gospel. Parliament ordered that the book be recalled and burned; see David Masson, *The Life of John Milton: Narrated in Connexion with the Political, Ecclesiastical and Literary History of His Time*, 7 vols (London, 1881–94), 4:423, 438–39.

not the case with Arianism, one can be confident in identifying an entire body of Socinian doctrines, the term was often used loosely in Milton's time as a broad brush to tar any theological opponent as unorthodox or atheistic; both terms were also used interchangeably to label any form of antitrinitarianism.

One can confidently label as Socinian seventeenth-century individuals who accepted the teachings of Sozzini and the tenets of the *Racovian Catechism*. By this standard, neither Milton nor Newton was a Socinian, although in different ways they shared significant theological views with Socinians. One can, however, label Milton and Newton Arians, not because they shared some putative but unavailable suite of opinions of the historical Arius, but because they shared the one view of the historical Arius that had come by the seventeenth century to be synonymous with his name, the view that the Son was neither eternal nor co-essential with the Father, who did not share supreme Godhead with the Son and the Spirit. Michael Lieb had some reason to caution against assuming that those who shared these positions are Arians in the strictest sense: "I wish to dispel the tendency, particularly among Miltonists, to regard 'Arianism' as a monolithic phenomenon with a well-defined set of doctrines clearly and indisputably trace-able to the thought and writings of a particular individual."[7] Rowan Williams, as Lieb observes, has suggested that "The time has probably come to relegate the term 'Arianism'...to oblivion."[8] But whatever full set of beliefs the historical Arius might have held, Arianism meant something particular to Milton's and Newton's contemporaries, as a label for a set of beliefs that they shared. There is an irony here. One can label the two thinkers Arian because they shared *one* identifiable position with Arius; one cannot label them Socinian because neither accepted *all* the positions of Sozzini and the Polish Brethren. I will thus speak of their affinities with Socinianism and of their identification as, in seventeenth-century terms, Arians.

Arianism

While there was in the latter decades of the twentieth century vigorous debate over whether Milton's writings can be characterized as Arian, for reasons I will outline below nearly all Milton scholars now acknowledge his Arianism.[9] Milton explicitly opposes the attribution to the Son of the divine attributes of co-eternity and self-existence; he insists that the Son was begotten in time, writing that

[7] Michael Lieb, "Milton and 'Arianism,'" *Religion & Literature* 32, no. 2 (2000): 197–220 (200).

[8] Rowan Williams, cited in Maurice Wiles, *The Archetypal Heresy: Arianism through the Centuries* (Oxford: Oxford University Press, 2001), 4.

[9] The authoritative account of Milton's Arianism is Michael Bauman's *Milton's Arianism* (Frankfurt: Peter Lang, 1986). John Rumrich clearly lays out the implications of the poet's Arianism in "Milton's Arianism: Why It Matters," in *Milton and Heresy*, ed. Stephen B. Dobranski and John Rumrich (Cambridge: Cambridge University Press, 1998), 75–92.

"God begot the Son by his decree, and likewise in time; for the decree must have preceded the decree's execution, as the added word 'today' sufficiently declare. So nowhere do I find what passage of the scriptures those who posit the Son's eternal generation protect themselves with" (*DDC* I.5; *OCW* 8.1:133). Milton is adamant that the Son does not share the Father's essence, and he is caustic in dismissing the pernicious importation into the theology of the godhead of irrelevant and nonsensical metaphysical speculation:

> If therefore the Son...does not also have an essence proper to himself, but has the Father's, then assuredly by their [orthodox trinitarians'] opinion he is rendered either not an *ens*, or else plainly the same *ens* as the Father; a teaching which indeed utterly overturns the Christian religion. But the answer which is usually given—that there can indeed be only one person of one finite essence, whereas there can be several persons of one infinite essence—is laughable; for because of the very fact that it is an infinite essence, all the more can it be of only one person....
>
> (if we are willing to dismiss metaphysical absurdities) substantial essence and subsistence are the same thing. (*DDC* I.5; *OCW* 8.1:153, 201)

For Milton, orthodox trinitarian claims amount to a logical contradiction and a violation of the most basic rule of arithmetic: "unless the reckoning, of one and two at least, were always the same with God and humankind, in vain would God so often have thundered out that first commandment—that he is the one and only God—if another besides be said to exist who, himself too, ought to be considered the one God" (*DDC* I.5; *OCW* 8.1:137).

Not surprisingly, there has been resistance to the proposition that Milton espoused an antitrinitarian view of the Son of God, and in particular that his great Christian epic diverges from orthodoxy. Some scholars have been unwilling to concede that *Paradise Lost* contained what Maurice Wiles has termed "the archetypal heresy."[10] C. S. Lewis argued that mining the poem for the heresies of the treatise "ignores the fact that this poem was accepted by many generations of acute readers well versed in theology."[11] But some readers detected, before the unearthing of *De Doctrina Christiana*, the unorthodoxy that hides in plain sight in *Paradise Lost*. Charles Leslie in 1698 criticized Milton for "mak[ing] the Angels ignorant of the blessed Trinity; and...tak[ing] it ill to acknowledge him for their King whom they had always ador'd as their God; or ask if the Son...had not been begotten till that day."[12] The Jonathan Richardsons, father and son, refer in 1734

[10] Wiles, *Archetypal Heresy*.

[11] C. S. Lewis, *A Preface to "Paradise Lost,"* (New York: Oxford University Press, 1942), 82.

[12] Charles Leslie, 'Preface to *The History of Sin and Heresy*', excerpted in John T. Shawcross, *Milton: The Critical Heritage* (London: Routledge & Kegan Paul, 1970), 117. On this and the following example, see Rumrich, "Milton's Arianism," 76, and Bauman, *Milton's Arianism*, 277–80, 285–87.

to a "Conjecture that Some have made; I mean that Milton was an *Arian*, and This is built on Certain Passages of *Par. Lost*."[13] "Theophilus," in a letter to the *Gentlemen's Magazine* in March, 1738, charges that "Milton has certainly adopted the *Arian* Principle into his *Paradise Lost*," a poem that "corrupt[s] our Notions of the most sacred Things."[14]

From the time of the discovery of the unpublished Arian manuscript of *De Doctrina Christiana* two centuries ago, there have been two strategies to insulate Milton from the charge of Arianism: tendentious reading of the treatise and questioning its attribution to Milton. When Bishop Thomas Burgess in 1829 disputed Milton's authorship of the manuscript, he announced his motivation in his book's title, *Milton Not the Author of the Lately-discovered Arian Work De Doctrina.* Burgess' reading of the treatise as denying the coequal and coeternal deity of the Son of God, while correct, led to his erroneous conclusion that Milton simply could not have written it.[15]

When Burgess did not carry the day on the question of authorship, the preferred strategy for separating Milton from Arianism shifted to the argument that neither *Paradise* Lost nor *De Doctrina Christiana* are Arian works. William B. Hunter and J. H. Adamson published essays denying Milton's Arianism in quick succession in 1959 and 1960, followed a few years later by C. A. Patrides.[16] They argued that Milton's Son of God was not Arian but "subordinationist," acknowledging that Milton viewed the Son as inferior to the Father but that, given the poet's view that the Son shared the substance of the Father, Milton remained trinitarian. John Rumrich dispatched that argument, by pointing out that for Milton all creation shares the Father's substance.[17] With that avenue closed, Hunter turned his attention next to resuscitating Bishop Burgess' argument against Milton's authorship of the treatise with ultimately no more success in the twentieth century than Burgess than in the nineteenth, though for a time it did require from anyone using *De Doctrina Christiana* a tedious rehearsal of the evidence for Milton's connection to the manuscript.[18]

[13] Jonathan Richardson, Father and Son, *Explanatory Notes and Remarks on Milton's Paradise Lost*, excerpted in John T. Shawcross, *Milton: The Critical Heritage, Volume II, 1732–1801* (1972; rpt. London: Routledge, 1995), 84.

[14] Letter excerpted in Shawcross, *Critical Heritage II*, 93.

[15] Throughout the work Burgess betrays a predisposition to view Milton as consistently orthodox throughout his life. He writes, for example, of an apparently Arian passage in the indisputably Miltonic *Artis Logicae* that "My view of Milton's religious opinions leads me to understand it in a sense expressive of Orthodoxy, and worthy of the most Orthodox believer." Thomas Burgess, *Milton Not the Author of the Lately-discovered Arian Work De Doctrina* (London: Thomas Brettell, 1829), 120. Burgess comments on the *Art of Logic* I.7 (*CPW* 8:233–34).

[16] Hunter, Patrides, and Adamson joined forces to publish *Bright Essence: Studies in Milton's Theology* (Salt Lake City: University of Utah Press, 1971), which contains their essays denying Milton's Arianism.

[17] John Rumrich, *Milton Unbound: Controversy and Reinterpretation* (Cambridge: Cambridge University Press, 1996), 43.

[18] The renewed debate was settled to the satisfaction of most Milton scholars by the appearance of Gordon Campbell's, Thomas N. Corns', John K. Hale's, and Fiona J. Tweedie's *Milton and the Manuscript of "De Doctrina Christiana,"* which demonstrated that Milton, in terms of the conventions

The ideologically interested resistance to the evidences of Milton's Arianism and his authorship of *De Doctrina Christiana* seemed at least somewhat plausible because of a dearth of professing Arians in the mid-seventeenth century. Hunter pointed to the absence of intellectually respectable Arians among Milton's contemporaries as evidence that the Milton could not have written *De Doctrina Christiana*, a work that Hunter belatedly acknowledged as Arian: "Elsewhere in England, antitrinitarianism was indeed not dead in Milton's own day. Masson has shown that such men as Thomas Webb, Paul Best, John Biddle, Samuel Richardson, and John Goodwin were active in supporting this thesis during the mid-1640s. But these are relatively unimportant writers who never achieved any real intellectual leadership."[19] Whatever one might say of a dubious assumption that Milton could not have registered an independent doctrinal view, Hunter's argument exemplifies how, until recently, the temporal perspective of Milton scholarship has been largely retrospective, treating Milton as the last of the Renaissance poets, as opposed to a writer of the Restoration or the long eighteenth century. The landscape changes when one considers Milton immediate successors. Had Hunter looked to the decades immediately following rather than before the time when Milton composed *De Doctrina Christiana*, he would have discovered that two figures who did achieve "real intellectual leadership," Newton and John Locke, were antitrinitarians, and that the young Newton was working his way toward an Arian conception of the Son of God not much later than his older contemporary Milton. Wiles has demonstrated that Arianism, though more or less invisible during Milton's time, flourished after 1688 in England, nurtured by the greater freedom of discussion of religion after the accession of William and Mary.[20] The Arian revival in the last decade of the seventeenth century and the first decade of the next was coaxed into the open by the passage of the Toleration Act of 1689, despite the fact that non-trinitarians were excluded from toleration. Newton would become a leading if guarded figure in an Arian and Socinian circle in the last years of the seventeenth century.

And though it was in the two decades beginning in 1690 that antitrinitarian thought—much of it Arian and Socinian—proliferated widely, Newton arrived at his own Arian commitment much earlier, all but certainly while Milton was still alive. In a manuscript dated between 1672 and 1675, Newton laid out twelve Arian

of authorship governing early modern systematic theology, which allowed for a great deal of unacknowledged quotation and paraphrase along with original content, is responsible for *De Doctrina Christiana*. I find unconvincing recent efforts to revive the debate, David V. Urban, "The Increasing Distance between *De Doctrina Christiana* and Milton's Poetry: An Answer to John K. Hale," *Connotations* 32 (2023): 1–10, and James M. Clawson and Hugh F. Wilson, "*De Doctrina Christiana* and Milton's Canonical Works: Revisiting the Authorship Question," *Renaissance and Reformation* 44 (2021): 151–98. The latter's stylometric arguments do not take into sufficient account the fact that much of the treatise is paraphrased and copied from contemporary sources.

19 Hunter, "Milton's Arianism Reconsidered," in *Bright Essence*, 33; Hunter cites Masson's *Life of John Milton*, 3:157 ff.

20 Wiles, *Archetypal Heresy*, 67.

principles of Christology, most of which parallel Milton's Christology. Newton begins with three observations on the meaning of the name God:

1. The {word} God is no where in the scriptures used to signify more then one of the thre persons at once.
2. The word God put absolutly without particular restriction to the Son or Holy ghost doth always signify the Father from one end of the scriptures to the other.
3. When ever it is said in the scriptures that there is but one God, it is meant of the Father[21]

Milton likewise argues that the word God used without qualification in Scripture refers to the Father alone, as he does in explicating 1 Corinthians 8:4: "there is none other God but one." Milton writes, "Here, *there is no other* (or second) *God but one* rules out not only a second essence but also any second person whatever, for distinctly, v. 6: *that one is the Father.* So there is no other person but one" (*DDC* I.5; *OCW* 8.1:141). Milton's conclusion is the same as Newton's: "As regards the name of God, then, where both Father and Son are mentioned together, the name is consistently attributed to the Father alone" (*DDC* I.5; *OCW* 8.1:153).

Newton resembles Milton as well in claiming that divine attributes belong essentially to the Father alone, and that the Father may in part share with or delegate to the Son those attributes. While the Father, according to Milton, can and does impart "to the Son as much divine nature as he wished" (*DDC* I.5; *OCW* 8.1:135), he does not and cannot make the Son, begotten in time and not sharing an essence with the Father, a co-equal, co-eternal God. The tenth of Newton's twelve Arian principles from the early 1670s echoes this idea:

> 10. It is a proper epithete of the father to be called almighty. For by God almighty we always understand the Father. yet this is not to limit the power of the Son, For he doth whatsoever he seeth the Father {do}, but to acknowledg that all power is originally in the Father & & [sic] that the son hath no power in him but what he derives from the father for he professes that of himself he can do nothing.

One of these powers is the power of creation. Newton's almighty God the Father, like Milton's, creates the universe by means of the Son, who acts as the chief or guiding alchemical agent of creation.[22]

[21] "Miscellaneous notes and extracts on the Temple, the Fathers, prophecy, Church history, doctrinal issues, etc.," Yahuda Ms 14, fol. 25^{r}; http://www.newtonproject.ox.ac.uk/view/texts/normalized/THEM00057, accessed January 29, 2024. This list of doctrinal points is quoted in Westfall, *Never at Rest*, 315–16; I follow Westfall's dating of the manuscript.

[22] Betty Jo Teeter Dobbs, *The Janus Faces of Genius: The Role of Alchemy in Newton's Thought* (Cambridge: Cambridge University Press, 1991), 83.

Newton agrees with Milton in ascribing omnipresence to the Father alone. In another manuscript set of articles of faith, an aged Newton would write, "The father is immoveable no place being capable of becoming emptier or fuller of him then it is by the eternal necessity of nature: all other beings are moveable from place to place."[23] Milton acknowledges that Matthew 18:20 ("where two or three are gathered in my name, there I am in the midst of them") shows that the Son can be on Earth and in Heaven at once, but he denies that "these words encompass an infinity of omnipresence)" (*DDC* I.5; *OCW* 8.1:205). Milton affirms the restriction of omnipresence to the Father alone in *Paradise Lost* as well. Returning to Heaven from serving as the Father's instrument in creation, "The Filial Power arrived, and sat him down / With his great Father, for he [i.e., the Father] also went / Invisible, yet stayed, (such privilege / Hath omnipresence)" (7.587–90).[24]

Both Milton and Newton specify that language indicating the unity of the Father and the Son (e.g., John 10:30, "I and the father are one") does not point to shared essence but rather agreement of wills. According to Milton, "The Father and Son are definitely not one essentially, for he himself has said the opposite…: *my Father… is greater than all….* [H]e declares that he and the Father are one in the same way that we are one with him, and that is definitely not in essence, but in love, in communion, in agreement, in charity" (*DDC* I.5; *OCW* 8.1.147). In the last of the early list of twelve Arian principles quoted above Newton agrees: "12 The union between him & the father he interprets to be like that of the saints one with another. That is in agreement of will & counsil."

Milton and Newton likewise agree that the Son is not omniscient. Principle 7 in Newton's early manuscript reads, "7 The Son acknowl{ed}g{e}th the original præscience of all future things to be in the father onely." Milton cites a quiver of scriptural verses in which the Son testifies that what he knows he has learned from the Father, and concludes, "But not even the son knows absolutely everything, for there are things which the father has kept secret to himself alone" (*DDC* I.5; *OCW* 8.1.207).[25] Not surprisingly, given that both Milton and Newton build their Christologies by applying reason to Scripture alone, they most often quote the same proof texts. On omniscience being the Father's alone, for example, both cite Matthew 24:36 and Mark 13:32, in which the Son of God speaks of the knowledge of when the world will end. Matthew writes "of that

[23] "Twelve Articles on Religion," Keynes Ms. 8, King's College, Cambridge, UK. https://www.newtonproject.ox.ac.uk/view/texts/normalized/THEM00008, accessed February 4, 2014). When Milton lays out the divine attributes delegated provisionally and incompletely to the Son, he mentions only four—omnipresence, omniscience, authority, and omnipotence (*DDC* I.5; *OCW* 8.1.204–7)—whereas in the chapter on God, he lists fifteen divine attributes: He is true God, a spirit, infinite, eternal, immutable, incorruptible, omnipresent, omnipotent, one, living, omniscient, holy, blessed, faithful, and just (*DDC* I.2; *OCW* 8.1:34–47).

[24] Bauman demonstrates conclusively, if one needs demonstration, that Milton refers to the Father rather than the Son in the latter half of the quoted passage; see *Milton's Arianism*, 252–56.

[25] "That the divine attributes [including 'omniscience' and 'supreme dominion'] also belong to the Father alone is also taught by the Son, to the exclusion even of himself" (*DDC* I.5; *OCW* 8.1:155).

day and hour knoweth no man, no, not the angels of heaven, but my Father only," while according to Mark the Son adds himself to the list of those who don't know: "not the angels which are in heaven, neither the Son, but the Father." Milton notes that Mark makes the point "more clearly/*clarius*" (*DDC* I.5; *OCW* 8.1:155, 154). Newton demonstrates his more exhaustive philological knowledge: Origen, Chrysostom, Theophylact, Hilary & Augustine include "neither the Son" in their commentaries on Matthew 24, which lacks the phrase in modern Bibles, demonstrating, as does its inclusion in the verse in Mark, that the text of Matthew has been corrupted.[26]

Most significantly for their Arian position, Newton, like Milton before him in the unpublished theological treatise, disputes the authenticity of two Scriptural verses that served as crucial foundations for trinitarian belief: 1 John 5:7 (the so-called Johannine comma) and 1 Timothy 3:16. They call for the excision of 1 John 5:7—"For there are three that bear record in heaven, the Father, the Word, and the Holy Ghost: and these three are one"[27]—arguing that the verse is a later interpolation missing from the earliest manuscripts. Each notes the absence of the verse in Syriac, Ethiopic, Arabic, and Greek codices, and each points to the authority of Erasmus for disregarding the apparently trinitarian verse.[28] Characteristically, Newton's philological evidence is more exhaustive than Milton's. Milton writes that "in the Syriac and the two remaining versions of the Orientals (the Arabic and the Ethiopic of course), and in most ancient Greek codices, this verse is not to be found—and it is phrased with remarkably variety in those where it is found!" Newton adds to this list an impressive additional array:

> Tis wanting also in other ancient versions, as in the Egyptian Arabic published in Walton's Polyglott, in the Armenian used ever since Chrysostom's age by the Armenian nations & in the Illyrican of Cyrillus used in Rascia, Bulgaria Moldavia Ruscia Muscovy & other countries which use the Sclavonick tongue. In a copy of this version printed at Ostrobe in Volkinia in the year 1581, I have seen it wanting, & one Camillus relates the same thing out of ancient manuscripts of this Version seen by him. Father Simon notes it wanting also in a certain Version of the French Church, which (saith he) is at least a thousand years old & which was published by F. {Mabillon} a Benedictine Monck.[29]

[26] "An historical account of two notable corruptions of Scripture, in a Letter to a Friend," Ms. 361(4), New College Library, Oxford, 91^{r}; https://www.newtonproject.ox.ac.uk/view/texts/normalized/THEM00264, accessed December 19, 2023. Milton *DDC* I.5, *OCW* 8.1:154–55.

[27] I quote the Oxford translation of Milton's citation of the Junius-Tremellius Bible (*OCW* 8.1:149, *CPW* 6:221). Newton in his letter quotes the King James version, "Two Notable Corruptions," fol. 19r; https://www.newtonproject.ox.ac.uk/view/texts/normalized/THEM00261, accessed June 8, 2022.

[28] *DDC* I.5, *OCW* 8.1: 149; Newton's "Two Notable Corruptions," 1^{r}.

[29] Newton, "Two Notable Corruptions," 6^{r}–7^{r}.

Newton's exposition of his Arian Christology is more thorough than Milton's both in its critical review of Church Fathers and its voluminous and painstaking demonstration of the corruption of Scripture.[30]

Both flag a rogue variant in 1 Timothy 3:16, which reads in part "great is the mystery of godliness: God was manifest in the flesh." According to Milton, "that passage in 1 Timothy 3:16, *God made manifest in the flesh*, suffers ... seriously from an uncertain reading, for in Ambrose and the Vetus Interpres, along with a great consensus of ancient exemplars, one does not read 'God' here, as Erasmus again witnesses" (*DDC* I.5; *OCW* 8.1.177). Newton devotes to this point page after page of dense philological analysis, going well beyond Erasmus, and laying out the mechanism of the corruption:

> by changing ***O*** in ***OΣ*** & both into ***ΘΣ*** (the abbreviation of **Θεὸς**) they now read Great is the mystery of godlinesse **God** manifested in the flesh: whereas all the Churches for the first four or five hundred years, & the authors of all the ancient Versions, Ierome as well as the rest, read, Great is the mystery of godliness **which was** manifested in the flesh. For this is the common reading of the Ethiopic, Syriac & Latine Versions to this day, Ierome's manuscripts having given him no occasion to correct the old Vulgar Latine in this place. Grotius adds the Arabic, but the Egyptian Arabic Version has **Θεὸς**, & so has the above mentioned Sclavonian Version of Cyrillus.[31]

This is merely the opening of Newton's demonstration. He goes on to cite Tertullian, Cyprian, Alexander of Alexandria, Athanasius, Epiphanius, Basil, Gregory Nazianzen, Gregory of Nyssa, Chrysostom, Cyril of Jerusalem, Cyril of Alexandria, Cassian, Hilary, Jerome, Ambrose, Austin, Phœbadius, Victorinus Afer, Faustinus Diaconus, Pope Leo the great, Arnobius Junior, Cereatis, Vigilius Tapsensis, Fulgentius, and the Bishops of the Council of Sardica, all this on the first page of more than ten closely-written manuscript pages devoted to his proof. To read Newton's manuscripts is to be stunned by the torrent of references. Locke, the massive letter's recipient, justly testified to Newton's "great knowledge in the Scriptures, wherein I know few his equals."[32]

[30] Sharon Achinstein traces the evolution of Milton's scholarly practice prompted by the harsh attacks on the first edition of *The Doctrine and Discipline of Divorce* and on display in the second edition, as he now drew support for his controversial arguments from ancient Patristic and Jewish sources and from philological examination of Scripture; see her "High Enterprise: Milton and the Genres of Scholarship in the Divorce Tracts," in *Scholarly Milton*, ed. Thomas Festa and Kevin J. Donovan (Liverpool: Liverpool University Press, 2019). Nevertheless, Milton's citation of the Church Fathers and his tracing of variations in the texts of the Bible, in the divorce tracts or in *De Doctrina Christiana*, never matched the exhaustiveness of Newton's. The avalanche of source citations on display here and in explication of the prophetic texts discussed in Chapter 3 recall Newton's scouring of ancient texts for the *prisca theologia* and *prisca scientia* mentioned in the Introduction.

[31] Newton, "Two Notable Corruptions," 26ʳ.

[32] Letter from Locke to Peter King, April 30, 1703; quoted in Wiles, *Archetypal Heresy*, 78–79.

Given the lines of possible transmission outlined in the Introduction, it is not impossible that Milton could have influenced the Arian opinions of Newton and his circle late in the century; it is evident, in any event, that Milton settled early on ideas that would be adopted by Newton, Locke, and others.

What were the implications of shared Arianism for the thought of Milton and Newton? (The implications *for them*, given seventeenth-century blasphemy laws, were potentially fatal.) Newton was developing his Arian conception of the Son at the same time that he was developing his vitalist understanding of matter. This seems to have been the case with Milton as well. For both an Arian understanding of the Son underwrites a vision of the relation of rational creatures to creation and to God. But beneath this shared understanding lies a radical difference between Milton's and Newton's visions. I turn now to what is shared between and what fundamentally divides Milton and Newton.

As we saw in Chapter 1, the living matter of *Paradise Lost,* unlike the inert and passive matter metaphysicians as different as Descartes, Hobbes, and the Cambridge Platonists proposed, can act. If Hell is a "universe of death" (2.622), ours is a living universe, in which even ostensibly "inanimate" objects act with some degree of freedom. The "pure immortal elements" of Eden "Eject [man] tainted now, and purge him off / As a distemper, gross to air as gross" (11.50–53). If, in *Paradise Lost,* freedom goes all the way down to "inanimate" matter, it also goes all the way up to God the Father and the Son. Milton writes that the "Father begets [the Son] not by necessity of nature but by free will," and that the Father "is shown from the holy books to do all things absolutely freely, from the intention of his own will" (*DDC* I.5; *OCW* 8.1:133). The Father in *Paradise Lost* is "free / To act or not" (7.171–72). Like Father, like Son. Both are free. Milton's Arian Son, while on the one hand serving as the "substantial expression" of the Father and the "effectual might" wielded by the Father, is at the same time separate from the Father. While, as we have seen above, Milton observes in *De Doctrina Christiana* that the Son is one with the Father not "in essence, but in love, in communion, in agreement, in charity" (*DDC* I.5; *OCW* 8.1.147), he also insists "it is taught from many passages that Father and Son do not in a numerical sense have the same intelligence nor will" (*DDC* I.5; *OCW* 8.1:157). The greater space in both essence and will between Father and Son in Milton's antitrinitarian view translates into greater independence for the Son and real freedom of choice. As Milton asserts in the *De Doctrina Christiana*, the Son redeems believers with his blood, "paid in full [and] voluntarily" (*DDC* I.14; *OCW* 8.1:469).

"Voluntarily" means one thing for Calvin and another for Milton. For Calvin, we do *voluntarily* what a predestinating God ordains, in the sense that we do not resist the ordained action by, for example, accepting or rejecting faith. For Milton, a voluntary act is one that one chooses oneself. In *Paradise* Lost, when the Father calls for a volunteer to sacrifice himself to save humankind, "all the Heav'nly choir stood mute, / And silence was in Heav'n: on man's behalf / Patron or intercessor none appeared" (3.217–19). At that point "all mankind / Must have been lost" (3.222–23), if the Son had not stepped forward. Either this moment is as cynically stage-managed as Satan's volunteering in Book 2, or the Son's choice is free in the strong sense that he could have chosen not to offer himself. The Father confirms the latter alternative, proclaiming him "By merit more than birthright Son of God" (3.309).[33] The Son's separateness from the Father opens a space for free choice and for merit.[34] The Son becomes the originary exemplar of Christian liberty, the internal freedom that for Milton is an argument for and the condition of political liberty. The freedom of the Christian is inscribed for Milton in the delegated divinity of the Son of God. This universal freedom is at the heart of Milton's theodicy.

An Arian Son is no less central to Newton's system, but the Son drives Newton's system in a direction different from Milton's. Where Milton is intent on demonstrating freedom from the top to the bottom of the chain of being, Newton focuses on God's omnipotent control. Agreeing with Milton (and with virtually all Christians) that God himself is ineffable, Newton insisted that God must act through instruments. Newton argued also that the active principles inherent in matter, which he investigated through tireless alchemical experimentation, account for all movement that is not merely mechanical.

Newton was working his way toward Arianism at the same time that he was developing his alchemical cosmology. As we saw in the Chapter 2, Newton's God endowed the universe with vital "aethereall Spirits" that formed the universe, not merely its organic bodies but the grand celestial order, in a manner not possible for mechanical laws. To return the "Hypothesis of Light" letter, Newton there writes that the "frame of nature" may consist of

> certain æthereal spirits, or vapours, condensed...; and after condensation wrought into various forms; *at first by the immediate hand of the Creator;* and ever since by the power of nature; which, by virtue of the command, increase and multiply, become a complete imitator of the copies set her by the protoplast. Thus perhaps may all things be originated from æther.[35]

The "immediate hand of the Creator," for Newton as for Milton, is the hand of the Son of God, the instrument or mediator through whom the ineffable Father creates. To understand what might seem a contradiction in describing the Son as instrument as the "immediate hand" of the Father, one can look to Newton's understanding of the "unity of dominion" of the Father and the Son, as expressed,

[33] I return to this critical moment in the epic in the Conclusion, as I examine how Milton meets the challenge of translating Arian Christology into narrative.

[34] Rumrich, "Milton's Arianism," 85–86.

[35] Newton, "Hypothesis explaining the properties of light," in Thomas Birch, *The History of the Royal Society*, vol. 3 (London: 1757), 3. 250 (my emphasis). For a relevant discussion, see Westfall, *Never at Rest*, 304–10.

for example, in his "Drafts on the History of the Church": "the Son receiving all things from the father, being subject to him, executing his will, sitting in his throne & calling him his God, & so is but one God with the ffather as a king & his viceroy are but one king."[36] For Newton, the ineffable, invisible God acts through his Son.

In the passage from the "Hypothesis of Light" just quoted, God endows or informs the world with living spirits, the alchemical "protoplasts," which in turn form bodies and create organisms and material structures unachievable by merely mechanical laws. As the Father, the one and only God, works through his Son, with whom He shares dominion, the Son then endows the pre-creation chaos with the living, alchemical spirits, and thus becomes himself in a sense the apex of a pyramid of alchemical spirits. According to Dobbs, "As the Christ acted in his capacity of assistant to the Father to frame the cosmos in the beginning, and since the creation of the world prefigured the alchemical work, the active agent in alchemy is thus identified as the Logos—still acting as God's creative agent in the framing of the world of matter."[37]

For Newton, the Son is the obedient instrument of the one God, the Father, in enacting divine commands in creating, in establishing the laws governing creation, and in dealing with rational creatures. Given the "unity of dominion" Newton describes in the "Drafts on the History of the Church," and given his repeated emphasis on the absolute sway of divine omnipotence over not only salvation history but also the history of peoples and sovereigns, the Son acts not by an independent will but as an expression of the will of the Father. For Milton, on the other hand, the Son is an exemplar of the freedom that characterizes every part of the chain of being. Milton's insistence that, to return to the passage from *De Doctrina Christiana* above, "Father and Son do not in a numerical sense have the same intelligence nor will" (*DDC* I.5; *OCW* 8.1:157), means that his Son, while *certainly* an instrument of the Father in creation and in interaction with his creatures, is not *merely* an instrument, but possesses his own agency.

This crucial difference between Milton and Newton informs their opposed positions on what can seem a merely arcane Christological detail, the hypostatic union of God and man in Christ. While Newton believes that the non-eternal begotten Son of God did become incarnate, in the eighth and ninth principles of his early Arian manuscript he rejects the hypostatic union:

[36] Newton, Yahuda Ms. 15.7, f. 154^r; https://www.newtonproject.ox.ac.uk/view/texts/normalized/THEM00237, accessed January 29, 2024. This passage is quoted also in Dobbs, *Janus Faces of Genius*, 82, and James E. Force, "Newton's God of Dominion: The Unity of Newton's Theological, Scientific, and Political Thought," in *Essays on the Context, Nature, and Influence of Isaac Newton's Theology*, ed. Force and Richard H. Popkin (Dordrecht: Kluwer, 1990), 79.

[37] Dobbs, *Janus Faces of Genius*, 83.

> 8 There is no where made mention of a humane soul in our saviour besides the word, by the mediation of which the word should be incarnate. But the word it self was made flesh & took upon him the {form} of a servant.
>
> 9. It was the son of God which he sent into the {world} & not a humane soul that suffered for us. If {there} had been such a human soul in our Savour [sic] it would have bee{n} a thing of too great consequence to have been wholly omitted by the Apostles.[38]

Jesus Christ is here for Newton not the hypostatic union of God's Son begotten before creation with the human son of Mary, but a divine figure clothed in human flesh. While both Milton and Newton complain bitterly about the corruption of biblical interpretation by the importation of Greek metaphysics, Milton, unlike Newton, finds the hypostatic union in Scripture:

> The union of two natures in Christ was the mutual hypostatic union of two essences. For where a perfect substantial essence exists, there a hypostasis or subsistence cannot be absent, since they are plainly the same thing; so that from that mutual hypostatic union of the two natures or essences has been formed one Christ, one *ens*, one person....And indeed the fact of the matter is self-consistent enough to be so; *how* it is so, is not known, and it is assuredly best that what God wills to be unknown should not be known. (*DDC* I.14; *OCW* 8.1:481)

This is the one place in *De Doctrina Christiana* where Milton allows mystery to trump reason, and he does so here because he finds the doctrine unambiguously present in Scripture. He does not attempt to explain "how it is so," leaving this one point behind the veil of mystery, something he refuses to do with the Trinity. In this case it is best to leave unknown "what God will to be unknown."

Why does Milton choose this one place to allow mystery to trump reason, given that Scripture is not unambiguously clear on a hypostatic union of God and man in Jesus Christ?[39] If it were, should we not expect Newton to have accepted the hypostatic union as well? It is true that Milton is able to point in the English Bible to the seldom-used term "mystery" employed multiple times in relation to the relation of God and man in Christ.[40] As Rumrich has argued, Milton's apparently anomalous acceptance of the mysterious here is consistent with his monism and with his anti-Aristotelean position on matter and form; for Aristotle form is generic and matter is the principle of individuation, where for Milton the

[38] "Miscellaneous Notes and Extracts on the Temple, the Fathers, Prophecy, Church History, Doctrinal Issues, etc.," Yahuda Ms. 14, fol. 25, National Library of Israel, Jerusalem. http://www.newtonproject.ox.ac.uk/view/texts/normalized/THEM00057.

[39] I thank John Rogers for asking me this question.

[40] 1 Tm 3:16, Col 1:26–27, 2:2–3, 4:3, Eph 1:9–10, 3:4, 3:9.

opposite is the case. For Milton, as Rumrich argues, the "human matter in an individual human form constitutes the Anthropos of this paradoxically singular, dual being," the Theanthropos, the incarnate Son.[41] Newton reasons that, absent Scriptural evidence for the Incarnate Son's being given an individuating human form, that is, a human soul, there is no basis for belief that the Son had a human nature or constituted a human person. For Milton, given the undeniable formed human body of the man Jesus, born from the human mother Mary, he by definition had a human nature and was a human person, in addition to his divine nature and personhood.

Rumrich demonstrates the philosophical foundation for Milton's acceptance of the one dogma at least apparently at odds with logic. At the same time, I propose that the hypostatic union is congenial to Milton because it fits the leveling impulse, the readiness to downplay differences between the Son and the rational creature. The Son, like us, is free to choose, and his merit consists in his free obedience to the Father. Despite Milton's acknowledgment of the theology of atonement, the Son who interests him is not the crucified Son but the Son who freely chooses to give himself for humankind in obedience to the Father.[42] In reading, the mysterious hypostatic union in Scripture Milton finds an endorsement of his intuition that human merit and divine merit are continuous; that union confirms his sense that we are, or at least he is, like the Son of God.

Newton, on the other hand, held that "It was the son of God which he sent into the {world} & not a humane soul that suffered for us." It is the Word or Son, the immediate instrument of the Father, who does the work of reconciling the Father with humankind, not a being containing both the divine and the human. Newton is an Arian because he needs a mediator between an absolutely ineffable and infinitely distant God and the created world, including human beings. That mediator, begotten rather than made by the Father, acts not for himself but as God's instrument. Newton's rejection of the hypostatic union fits his emphasis on the distance between God and his begotten Son, on the one hand, and his creatures, on the other, even as he insists on the unique status of the Father as the one God. Milton's embrace of the hypostatic union despite his allergy to mystery fits his sense that in some important ways human beings and the Son of God resemble each other. Newton separates the divine and the human. The Son becomes an extra step, a quasi-divinity, between God and man. In Milton, God and man come together in the Son, and the distance is reduced.

[41] John Rumrich, "Milton's Theanthropos: The Body of Christ in Paradise Regained," *Milton Studies* 42 (2003): 55.

[42] See John Rogers' "Milton and the Heretical Priesthood of Christ," in *Heresy, Literature and Politics in Early Modern England*, ed. David Loewenstein and John Marshall (Cambridge: Cambridge University Press, 2006), 203–20.

Socinian Leanings

In agreeing that the Son is neither co-essential nor co-eternal with the Father and that he was the instrument through whom the Father created, Milton and Newton fit the criteria for Arianism as understood in their time. When it comes to possible connections to Socinian teachings, the case is more complicated, partly for the reason outlined at the beginning of this chapter: Socinianism embraced a large body of doctrines in Christology, metaphysics, soteriology, and sacramental theology, and both Milton and Newton agreed with Socinians on some points and disagreed on others. It is also the case that the thinking of both evolved over time and that Newton was not always consistent, particularly when it came to Christology.

Stephen Snobelen, a leading scholar of Newton's religious views, concludes that "Although formally Newton's Christology was Arian, he appears to have adhered to a practical Socinian Christology. Put another way, Newton's Christology stood on the Socinian side of Arianism rather than the orthodox side."[43] Wiles agrees: "though one might describe his [Newton's] religious position as predominantly Socinian, there is no doubt that his overall theological position is Arian rather than Socinian."[44] For students of Milton, these judgments will have a familiar ring, as recent studies have pointed toward Socinian elements in Milton's thought. Hugh MacCallum, John Rogers, Martin Dzelzainis, Nigel Smith, and Dustin Stewart have explored affinities between Milton's writing and Socinianism.[45] As noted above, while it would not be accurate to call either Milton or Newton *a* Socinian, they *did* adopt overlapping sets of Socinian positions.

Why, I will ask, did both Milton and Newton find themselves drawn to distinctively Socinian teachings, and what does this shared attraction tell us about them and about the intellectual culture of late seventeenth-century England? Milton, as I argue throughout this book, was an early adopter of views associated with the Newton circle in the 1690s. Milton and Newton, poet and natural philosopher, shared a markedly rationalist approach to scriptural interpretation and theology, and each was allergic to doctrinal mystery unless unambiguously found in the Bible, positions congenial to Socinians. Each, at least in part as a result, was

[43] Stephen D. Snobelen, "Isaac Newton, Socinianism, and 'the One Supreme God,'" in *Socinianism and Arminianism: Antitrinitarians, Calvinists, and Cultural Exchange in Seventeenth-Century Europe*, ed. Martin Mulsow and John Rohls (Leiden: Brill, 2005), 262.

[44] Wiles, *Archetypal Heresy*, 84.

[45] Hugh MacCallum, *Milton and the Sons of God: The Divine Image in Milton's Epic Poetry* (Toronto: University of Toronto Press, 1986); Rogers, "Milton and the Heretical Priesthood," 201–20; Martin Dzelzainis, "Milton and Antitrinitarianism," in *Milton & Toleration*, eds Sharon Achinstein and Elizabeth Sauer (Oxford: Oxford University Press, 2007), 171–85; Nigel Smith, "'And if God was one of us': Paul Best, John Biddle, and Anti-Trinitarian Heresy in Seventeenth-century England," in *Heresy, Literature, and Politics*, 160–84; and Dustin Stewart, "Paradise Regained on Socinian Time," *Religion and Literature*, 45 (2013): 180–92.

misread as supporting deist views. Milton found a champion in the deist John Toland, and Newton's physics was viewed, however inaccurately, as grounding a mechanist view of the universe as the clockwork of a deistic God.[46] These giants of worldmaking, in poetry and natural philosophy respectively, were supremely learned and confident interpreters of the Bible who independently found their way toward heterodox views. With Socinians and with other sectarians they defended believer's baptism, toleration, and the separation of church and state. They agreed with Socinian attacks on the corruption of Scripture by trinitarians; with Socinian insistence on the union of the Father and Son as one of will rather than essence; and with Socinian defenses of mortalism. Their self-understanding as prophetic members of a faithful remnant has Socinian overtones. At the same time, their divergences from each other and from Socinian tenets are just as instructive. They differed on the Socinian tenet of the freedom of the will and the identification of dominion as God's defining attribute, with Milton defending the former and Newton the latter, choices reflecting their commitment to creaturely liberty on one hand and divine omnipotence on the other. On the nature(s) of Christ, they disagreed with Socinians and, as we have seen above, with each other.

It can be challenging to sort out Arian and Socinian relations in the early modern period, when the two were often confused. The salient divergences of Socinian from Arian Christologies involve (1) the temporal location of the Son of God's origin—with Socinians claiming that the Son of God was begotten not before the creation of the universe but when conceived in the virgin Mary by the Father's Spirit—and (2) the nature of the Son—with Socinians claiming that the Son's nature was not dual but wholly human. Despite this crucial difference, the label of Socinian became a rough shorthand for any antitrinitarian view. In 1704, for example, John Dennis suggested darkly that "Milton was a little tainted with Socinianism, for by the first Verse [*Paradise Lost* 3.383–84, 'Thee next they sang of all creation first, / Begotten Son'] 'tis evident, that he look't upon the Son of God as a Created Being."[47] The position in these lines, in which the angels acknowledge the Son as the being begotten prior the rest of creation, is Arian rather than Socinian, as is the claim a few lines later: "He [the Father] Heav'n of Heav'ns and all the powers therein / By thee [the Son] created" (3.390–91).

It should be noted as well that, not surprisingly, Milton's and Newton's thinking changed over time. The young Milton was an orthodox trinitarian. His identifiably Arian statements come in his mature *De Doctrina Christiana* and *Paradise Lost*, and markedly Socinian sympathies emerge late in *Paradise Regained* and *Of*

[46] See James E. Force, "The Newtonians and Deism," in *Essays on the Context, Nature, and Influence of Isaac Newton's Theology*, ed. James E. Force and Richard H. Popkin (Dordrecht: Kluwer Academic Publishing, 1990), 43–73.

[47] John Dennis, *The Grounds of Criticism in Poetry* (London, 1704), 36. The passage is cited by Rumrich, "Milton's Arianism," 76.

True Religion, although Rogers makes a plausible case for a Socinianizing cast to Milton's narration of the Son's exaltation in *Paradise Lost*.[48] Newton is a tougher case, as he wrote on both sides of central points of contention among Arian, Socinian, and orthodox beliefs. In a manner not unlike the natural philosopher Newton, who both counseled against offering hypotheses on the cause of gravity and offered varying hypotheses on the same, the theological Newton both insisted that the Son of God was preexistent and suggested that he may have come into being when conceived in Mary. He apparently considered the question of when the Son of God came into existence as an instance of the "strong meats," the *adiaphora*, as opposed to the "milk for babes" containing the fundamental truths of Christianity. It was one of the questions about which, according to Newton, one might disagree without heresy, falling out of communion, or endangering one's soul.

Like Milton, Newton was suspected of Socinian leanings by early readers. Ten years after Dennis' comment on Milton, John Edwards in 1714 acidly observed that the idea that God, in the words of Isaac Newton's protegé Samuel Clarke, "*is a Relative Word, and a Word of Dominion of Power*," is taken directly from the Socinian Johannes Crellius' *De Deo et eius attributis* (1630). The words quoted from Clarke echo the Latin of the General Scholium in the 1713 edition of the *Principia Mathematica*, published the previous year. Clarke and Newton, Edwards implies, are tainted by Socinianism. Edwards, after quoting at length the offending passage in the General Scholium, goes on to ask, "Who wou'd have thought, that such wild Jargon as this, could come from the Pen of so great a man, and of so subact [i.e., disciplined] a Judgment as Sir *Isaac Newton*, who hath justly merited the Applause of the learned World, for his admirable Efforts in *Natural Philosophy and Mathematics*."[49] As Edwards continues, we again see the interweaving of the Arian and Socinian strains:

> He tells us next, That the *Supreme God* (which is the Epithet that all the *Arians* and *Socinians* use, to distinguish the Father from the Son, who they hold to be an *Inferior* God) is *Lord God*, because of his *Dominion*, for 'tis this *Dominion that makes him a God*. But why is this *Attribute* chosen out before all the rest, to constitute a *Deity*? One wou'd think, that *Goodness*, *Holiness*, *Mercifulness* and *Benignity*, should have had the Precedence.[50]

[48] Rogers, "Milton and the Heretical Priesthood."

[49] John Edwards, *Some Brief Critical Remarks on Dr. Clarke's Last Papers* (London, 1714), "Postscript," 38; quoted from https://newtonprojectca.files.wordpress.com/2013/06/edwards-some-remarks-on-clarkes-last-papers-postscript-1714-a4.pdf, accessed August 8, 2025. Newton's passage, in Edwards' translation: "*God* is a *Relative* Word, and hath a Reference to *Servants*: And the *Deity* is the *Dominion* of God, not on his own Body, but on Servants. The *Supreme God* is an Eternal, Infinite, and absolutely Perfect Being, but a Being tho' never so Perfect, without Dominion, is not *Lord God*—It is the *Dominion* of a Spiritual Being that makes *a God*. It his Dominion be true, it makes a True God; if Supreme, it makes a Supreme God, if False and Counterfeit, it makes a False God."

[50] Edwards, *Some Brief Critical Remarks*, 40.

Unlike the case of John Dennis and the passage from *Paradise Lost*, in which an Arian position is labelled Socinian, Edwards does here identify a peculiarly Socinian tenet: dominion as the defining attribute of God.

Both Milton and Newton had access to Socinian texts. Milton certainly knew the *Racovian Catechism* by 1650.[51] Hugh MacCallum points to the likelihood that a still younger Milton, while stopping in Siena on his Grand Tour in October of 1638, just shy of his 30th birthday, would have had occasion to ask about the Sienese native Sozzini, whose influence was proliferating across Europe. In Paris in May or June of that year, Milton visited the celebrated scholar Hugo Grotius, who was suspected, however mistakenly, of Socinianism.[52] Moreover, in the years when Milton was working on his theological treatise and late poems, he had access to his friend Nathan Paget's library, with its large collection of Socinian texts, including twenty-one books by Socinius himself (more than by any other one author), the *Racovian Catechism,* and several works by Crellius, including two in English translation.[53]

Newton himself owned eight books by Socinians, including two by Socinius and two by Crellius. And Newton had access to the extensive collections of Socinian texts in the libraries of John Locke and Samuel Clarke; the latter owned two sets of the *Bibliotheca Fratrum Polonorum*, one of which may have been inherited from Newton himself.[54] Newton friend and subordinate at the Royal Mint, Hopton Haynes, was a Socinian, who testified to his superior's Christological Socinianism and who as early as 1701 introduced Newton to Zsigmond Pálfi, a Unitarian (or Socinian) from Transylvania.[55] Later Newton met with Crellius' grandson the Socinian minister Samuel Crell, who in 1726 enlisted Newton's help in publishing a commentary on the Gospel of John, and who indicated in a letter that Newton had personally assisted him when he was in difficulties in 1711.[56]

[51] While raising legitimate questions about the common view that Milton was technically the licenser who in 1650 approved the publication of the Socinian *Racovian Catechism*, Stephen B. Dobranski notes that Milton's Dutch acquaintance Leo van Aitzema recorded in March of 1652 that Milton acknowledged his role in the controversial book's publication. Dobranski adds that "It is... pleasing to speculate whether the incident of *The Racovian Catechism* served as a catalyst for some of Milton's still inchoate heretical opinions—just as 'the acute and distinct Arminius' had been 'perverted' when called upon to refute a heretical 'nameless discours.'" See Stephen B. Dobranski, "Licensing Milton's Heresy," in *Milton and Heresy*, 142–48 (quoted passage is on 148). Dobranski quotes Milton from the *Areopagitica* (*CPW* 2:519–20), where he wonders if Arminius was persuaded to abandon his orthodox position on Calvinist predestination by reading works he was tasked to refute.

[52] MacCallum, *Milton and the Sons of God*, 53. MacCallum notes as well that Milton, decades later in *De Doctrina Christiana*, cites and answers the trinitarian arguments of the anti-Socinian Placaeus of Saumur (MacCallum, 278, n. 76).

[53] See Christopher Hill, *Milton and the English Revolution*, Appendix 3 "Nathan Paget and His Library" (New York: Viking, 1977), 494. There is a copy of the library catalogue, *Bibliotheca midica viri clarissimi Nathanis Paget, M.D.*, though significantly blurred, on Early English Books Online, ProQuest doc. 2240936276.

[54] For Newton's access to Socinian texts, see Snobelen, "Isaac Newton, Socinianism, and 'the One Supreme God,'" 252–55, 296–98.

[55] Snobelen, "Isaac Newton, Socinianism, and 'the One Supreme God,'" 248, 260.

[56] Snobelen, "Isaac Newton, Socinianism, and 'the One Supreme God,'" 250.

Milton closes his last prose work, *Of True Religion, Hæresie, Schism, and Toleration* (1673), much as he had opened his chapter on the "Son of God" in the *De Doctrina Christiana*, with a plea for toleration of different theological positions:

> How shall we prove all things, which includes all opinions at least founded on Scripture, unless we not only tolerate them, but patiently hear them, and seriously read them? If he who thinks himself in the truth professes to have learnt it, not by implicit faith, but by attentive study of the Scriptures & full perswasion of heart, with what equity can he refuse to hear or read him, who demonstrates to have gained his knowledge by the same way?... We suffer the Idolatrous books of Papists, without this fear, to be sold & read as common as our own. Why not much rather of Anabaptists, Arians, Arminians, and Socinians? (*CPW* 8:436–37)

Milton here does not merely list opinions feared to be heretical against the *bête noir* of implicit Roman Catholic faith. He groups Socinian teaching with two positions—Arminianism and Arianism—that he clearly supports in *Paradise Lost* and *De Doctrina Christiana*, and with Anabaptism, whose defining opposition to infant baptism he shares.[57] Earlier in the same work, Milton is sterner toward orthodox Calvinist predestinarians than toward supposed heretics.

> The Calvinist is taxt with Predestination, and to make God the Author of sin; not with any dishonourable thought of God, but it may be over zealously asserting his absolute power, not without plea of Scripture. The Arian and Socinian are charg'd to dispute against the Trinity: they affirm to believe the Father, Son, and Holy Ghost, according to Scripture, and the Apostolic Creed; as for the terms of Trinity, Triniunity, Coessentiality, Tripersonality, and the like, they reject them as Scholastic Notions, not to be found in Scripture, which by a general Protestant Maxim is plain and perspicuous, abundantly to explain its own meaning in the properest words, belonging to so high a Matter and so necessary to be known; a mystery indeed in their Sophistic Subtilties, but in Scripture a plain Doctrin. (*CPW* 8:424–25)

While Michael Lieb rightly points out that Milton here is primarily concerned with arguing for adherence to a Protestant mode of scriptural interpretation, I would like to put pressure on his conclusion that "It would be a mistake...to invoke the foregoing passage as a basis for claiming that Milton was a Socinian, an Arian, or anything else that constitutes his catalog."[58] The passage from *Of*

[57] For a lucid and compelling reading of Milton's views on baptism, see Alison A. Chapman, "'To All Baptized': The Watery Dissent of *Paradise Regained*," *Milton Studies* 65 (2023): 276–99.

[58] Michael Lieb, *Theological Milton: Deity, Discourse and Heresy in the Miltonic Canon* (Pittsburgh: Duquesne University Press, 2006), 241. The quoted passage is from a chapter, "The Socinian Imperative,"

True Religion reveals not only hermeneutical principles but also barely submerged judgments on sectarian positions. Milton gives the Calvinist the tepid endorsement of being "not without plea of Scripture," while describing the Arian and Socinian as believing in Father, Son, and Spirit "according to Scripture, and the Apostolic Creed." (For both Milton and Newton, the Apostles' Creed was the true primitive creed of the Christian faith.) Milton in *De Doctrina Christiana* himself charges Calvinist teaching with making God the 'author of sin.' In denying that divine decree restricts rational creatures" free wills, he concludes,

> neither God's decree nor his prescience impedes free causes by necessity. Yet some people do strive, by fair means or foul, to oppose this doctrine, not hesitating to affirm that God is, in himself, the cause and author of sin [*non dubitant Deum asserverare causam per se peccati et authorem*]. Such men, if I did not think they spoke like this from error rather than wickedness, I would consider the most abandoned of all blasphemers; and if I should try to refute them, I should be doing the same as if I argued at great length that God is not the Devil. (*DDC* I.3; *OCW* 8.1.66–67)

Given that the term "Socinian" was at the time a brush to tar heretics and theological opponents, it may not be surprising that Milton couches his defense of the good faith of Socinian as obliquely as he does in a text the main purpose of which is to promote broad tolerance of a spectrum of Protestant denominations and sects. For the same reason, his all but explicit defense of Socinianism is significant.

Socinians agree with Arians that the "orthodox" understanding of the Trinity amounts to polytheism. The *Racovian Catechism* holds that trinitarian belief "may easily undermine and weaken the belief of one God, whilest one while it confesseth but one God, another while acknowledgeth three Persons, every one whereof is that one God."[59] Socinians found equivocation on one and three to be a violation of reason and a stumbling block in proselytizing other monotheists, who would baulk at the multiplication of gods. According to the *Racovian Catechism*, "it greatly hindreth Aliens from embracing the Christian Religion, whilest it delivereth such things as are opposite to divine Truth and sound reason."[60] To promulgate doctrine that violates the simple sense of Scripture as available to reason is to erect a barrier to belief. Milton's attack on trinitarian thinking as polytheistic is strongly implicit in his discussion of God's ninth attribute, that he is one. He quotes a series of verses from the Hebrew scriptures insisting on monotheism,

previously published in *Milton and the Grounds of Contention*, ed. Mark R. Kelly, Michael Lieb, and John T. Shawcross (Pittsburgh: Duquesne University Press, 2003), 234–83.

[59] *The Racovian Catechism* (Amsterdam, 1652), 24; Early English Books Online, ProQuest doc. 2248533629, accessed August 8, 2025.

[60] *Racovian Catechism*, 24–25.

before elaborating, "What could have been stated that was plainer, what more definite, what more adapted to popular perception and to everyday usage of speech, so that God's people might understand that God is one in number, one spirit, even as they understood that anything else was numerically one?" (*DDC* I.2; *OCW* 8.1:41).[61] Newton is more explicit on the equation of trinitarianism and polytheism. For Newton, the original Noachide religion was monotheistic, and the history of religions has been an oscillation between corruption into polytheism and idolatry and prophetic reform and restoration of monotheism. As we have seen, he viewed his own work as participating in that prophetic reform, and his demonstrations of the error of trinitarian doctrine as crucial to that reform. In his manuscript on "The Original of Religions" (early 1690s), he traces the genesis of corruption, polytheism, and idolatry to Egypt: "because Egypt was the oldest of kingdoms, its reasonable to beleive that the Egyptians were the first who worshipped such Gods, & by consequence that the 12 Gods were their ancestors."[62] Newton's tenacious, indefatigable combing of gentile chronicles and chronologies and aligning them with biblical chronology ultimately derive from and feed his conviction that human history is a story of the struggle between rare prophetic figures (among whom Newton included himself) and those who promoted descents into idolatry and polytheism.

Milton's and Newton's shared beliefs extend to some that were common among other seventeenth-century sectarians—insistence on believer's baptism and pleas for toleration and the separation of church and state—and others more distinctive to Socinians—a denial of the immortality of the soul, an attack on philosophical sophistication and multiplication of metaphysical terms, and a philologically informed emphasis on the corruption of Scripture by the insertion of verses absent in earlier texts. The first, the denial of the immortality of the soul, was nearly as explosive as the denial of the Trinity. Socinians argued that the natural state of human beings is mortality; immortal life is a gift of God to those who merit it by obedience and virtuous life, and eternal death is the reward of those who transgress. Their position is laid out in question and answer form in the *Racovian Catechism*:

Q. ...you said that this way, which leadeth to immortality, was discovered by God, I would fain know for what reason you said so?

A. Because as man by nature hath nothing common with immortality, so could he not by himself possibly know the way leading thereunto.

Q. Why hath man nothing common with immortality?

[61] On the same page Milton insists that this foundational principle is not revised in the Greek Testament.

[62] Yahuda Ms. 41, fol. 11ʳ; https://www.newtonproject.ox.ac.uk/view/texts/normalized/THEM00077, accessed December 19, 2023.

> A. Because he was at first formed of the earth, and therefore created mortall. Again, because he transgressed the commandements of God proposed to him, and so by the decree of God himself expressed in the commandement was necessarily subjected to eternall death.[63]

The Socinian Joachim Stegmann, in a text translated by leading English Socinian John Biddle and published in 1653, argues that in accepting the natural immortality of the soul Lutherans and Calvinists cannot answer erroneous Papist teaching, including the

> opinion where they hold that the dead live.... It will seem absurd, and indeed the thing itself is very absurd. Yet they believe it. For they suppose that the souls of men in the very moment wherein they are departed from their bodies by death are carried either to heaven and do there feel heavenly joy and possess all kinds of happiness, which God hath promised to his people; or to hell and are there tormented and excruciated with unquenchable fire. And this, as was said before, they attribute to the mere souls separated from the bodies, even before the Resurrection of the men themselves, that is to say, while they are yet dead.[64]

Like the Socinians, Milton finds no evidence of the natural immortality of the soul in Scripture. Along with Stegmann, Milton denies that death can involve the separation of the body, which dies, and the soul, which lives; death entails the complete loss of life:

> What is called **bodily death** is the privation or annihilation of life. For the separation of soul and body, as death is commonly defined, can in no way be death: for what part of a person, I ask you, dies at that rending apart? Surely not the soul? But those who so define death deny that. Surely not the body? Yet how is something that never lived by its own power said to die? That separation, therefore, should not be called a person's death. (*DDC* I.13, *OCW* 8.1:441)

Like Stegmann's, the logic of Milton's position rules out belief in any "intermediate state" of life between physical death and the resurrection of the body at the Last Judgment (*DDC* I.13; *OCW* 8.1:447). Given Milton's animist materialism, it would be surprising were he not to have espoused mortalist views. He writes in *De Doctrina Christiana* that

[63] *Racovian Catechism*, 10–11.

[64] Joachim Stegmann, *A Brief Inquiry Touching a Better Way than Is Commonly Made Use of to Refute Papists and Reduce Protestants to Certainty and Unity in Religion*, trans. John Biddle (London, 1653). Excerpted in *The Polish Brethren: Documentation of the History & Thought of Unitarianism in the Polish-Lithuanian Commonwealth and in the Diaspora, 1601–1685*, ed. and trans. George Huntston Williams, Harvard Theological Studies XXX, 2 vols, continuously paginated (Ann Arbor: Scholars Press, 1980), 2:363.

> the idea of man's spirit being separated from his body, so as to exist somewhere apart, entire and intelligent [*Separari autem,spiritum hominis à corpore, ita ut alicubi seorsim integer et intelligens existat*], is not only not found anywhere in holy scripture, but also plainly clashes with nature and reason....And furthermore, it is said of every kind of animate being in Gen. 1: 30: *in which is a living soul*; and 7: 22: *everything in whose nostrils is the breath of the spirit of life, of all those on dry land, died.* And yet it is not for that reason believed that the soul of beasts exists separately somewhere. (*DDC* I.7; *OCW* 8.1:302–5)

The fallen, despairing Adam in *Paradise Lost* comes to the same conclusion on different grounds: "It was but breath / Of life that sinned; what dies but what had life / And sin. The body properly hath neither. / All of me then shall die" (10.789–92).

Newton writes of the resurrection of the dead body and soul several times in his manuscripts:

> Daniel tells us that in the great tribulation (that out of which the Palmbearing multitude comes) Michael shal stand up the great Prince which is set over the people of the Iews & they shall be delivered every one that shall be found written in the Book & that at the same time many of those that sleep in the dust shall awake some to everlasting life & some to shame & everlasting contempt.[65]

The end of this passage is a nearly word for word quotation from the King James translation of Daniel 12.2, an important source texts for mortalists. Newton quotes the passage to the same effect at least two decades earlier, in the early 1680s, in the manuscript "Prophecies concerning Christs 2nd coming."[66] In both of these manuscripts, Newton refers again to the prophetic group that he considers himself a part of, the "Children of the Resurrection," those who will at the millennium be raised first and rule with the returned Christ before the general resurrection and last judgment. In the manuscript "Of the Woman & two horned Beast," Newton writes, "That the resurrection from the dead is called living again & therefore between death & the resurrection men do not live. [¶] That men are rewarded at Christs coming not before."[67]

[65] "The synchronisms of the three parts of the prophetick Interpretation," Yahuda Ms. 6, 15r; http://www. https://www.newtonproject.ox.ac.uk/view/texts/normalized/THEM00049, accessed January 29, 2024. This passage is from the early 1700s.

[66] Andrews University Ms. N47 HER, 20r; James White Library, Andrews University, Berrien Springs, Michigan.

[67] Yahuda MS 7.2e, f. 4v; https://www.newtonproject.ox.ac.uk/view/texts/normalized/THEM00424, accessed January 29, 2024. See also "Paradoxical Questions," fol. 54r–55r; On Newton's mortalism, see Stephen D. Snobelen, "Isaac Newton, Heretic: The Strategies of a Nicodemite," *British Journal for the History of Science* 32, no. 4 (1999): 387, and James E. Force, "The God of Abraham and Isaac (Newton)," in *The Books of Nature, and Scripture*, ed. J. E. Force and R. H. Popkin (Dordrecht: Kluwer Academic Publishing, 1994), 179–200.

As James E. Force points out, Calvin cites Daniel 12:2 not to demonstrate the mortality of the soul but in the course of his assertion of the soul's immortality.[68] In the *Institutes* III.xxv.6, headed "Resurrection of the flesh but immortality of the soul," Calvin dismisses "Some [who] have thought, as if the whole man were to die, that the souls would be resurrected with the bodies," adding that "there is nothing uncertain in the words of Daniel, 'And many of those who sleep in the dust of the earth shall awake, some to everlasting life, and some to shame and everlasting contempt.'"[69] What is certain for Calvin is that dead bodies at the end of time are reunited with immortal souls. Like Calvin, Wolleb, one of the sources followed in Milton's *De Doctrina Christiana*, cites Daniel 12:2 as a proof text for the revivification of the body prior to its reunion with its immortal soul.[70]

The conclusion that Calvin draws from Daniel 12, while certain and obvious to him, is certainly erroneous for Newton and for Milton as well. As Milton argues,

> Since a whole person is frequently said to consist of body, spirit, and soul—in whatever way we may finally think those parts should be distinguished from each other—I shall demonstrate that first the whole person, then every single part of him singly, is deprived of life.... For what could have been juster than that someone who had sinned as a whole person should die as a whole person? And most of all that part, namely, the soul or the spirit, which has been found to be the leading part in sin. (*DDC* I.13; *OCW* 8.1:443)

Milton characteristically ties his conclusion on the soul's mortality to the imperative of divine justice; if (and this is a real if) we can distinguish between body and soul (the manner of distinction the materialist Milton acknowledges as not self-evident), it would be anomalous and unjust if the part that decides to sin escaped mortality, owing to an ontological escape hatch.

As noted in Chapter 1, Milton denies that a putatively immaterial soul and an obviously material body are joined across an ontological divide. To speak of body, spirit, and soul is to speak of aspects or facets of one undivided substance: "whenever we speak of the body as though of a torso, then 'soul' means either the same as 'spirit,' or its less primary faculties: the vital or sensitive faculty, say; and so it is no less often distinguished from the spirit than from the body" (*DDC* I.7; *OCW* 8.1:303). Soul belongs no less to the body than to the spirit; for Milton the terms provide different ways of approaching an indivisible whole composed of Raphael's "one first matter" (*Paradise Lost* 5.472). Body and soul are not "generically different and [of] distinct natures,...the whole man is soul, and the soul is man;

[68] Force, "God of Abraham and Isaac," 186.

[69] John Calvin, *Institutes of the Christian Religion*, ed. John T. McNeill and trans. Ford Lewis Battles. Library of Christian Classics, 2 vols (Philadelphia: Westminster Press, 1960), III.xxv.6 and 7 (II.996, 999).

[70] Johannes Wollebius, *Compendium Theologiae*, I.xx 2.1 (*Reformed Dogmatics*, 181).

namely a body or substance which is individual, animated, sensitive, and rational" (*DDC* I.7; *OCW* 8.1:303).

Whatever Calvin might have said about the certainty of the immortality of the soul and the restriction of Daniel 12:2 to bodily death and resurrection, Milton enlists authoritative forerunners in support of his diametrically opposite position: "That all the Saints and the faithful, the patriarchs, the prophets, and the apostles were of this opinion is clear" (*DDC* I.13; *OCW* 8.1:443), a claim he backs up with a several pages of verses drawn from Hebrew Scriptures and the Greek Testament, including Daniel 12:2 (*DDC* I.13; *OCW* 8.1:445).

The erroneous though generally accepted belief in the natural immortality of the soul arose, in Newton's eyes, from the malign manipulation of his archenemy Athanasius. Newton writes in his manuscript "Paradoxical Questions concerning y[e] morals & actions of Athanasius and his followers,"

> The Christians of the first ages taught that at the day of judgment, Christ would sentence some immediately to heaven others to perpetual torments & others to be baptized with fire & shut up in prizon till they should pay the outermost farthing. What was the state of souls between death & the day of judgment the Greek Churches of those ages (so far as I can find) determined not till Athanasius in the life of Antony by relating how Antony saw the soul of Ammon ascend up to heaven, brought in an opinion that the souls of the blessed went immediately after death to heaven & by consequence that those of the wicked went immediately to hell & those of the middle degree to Purgatory. And because its absurd that men should be rewarded before they are judged therefore the Athanasians have feigned without any foundation in scripture that there is a double judgment, the first particular of every man at his death & the second general.[71]

Newton ties the doctrine of the natural immortality of the soul, and the entailed belief that individuals live between the death of the body and the Last Judgment, to Roman Catholic idolatry: "Athanasius teaches three things, that the saints understand what we say, that they intercede with God for us, & that in certain forms of words there are supernatural virtues: & these principles readily inferr saint-worship & charms."[72] Socinians also tie belief in natural immortality to Roman Catholic idolatry. In the Stegmann text translated by Biddle, cited above, we find the same connection: "Wherefore they believe in effect that the dead live, namely, in the same manner that they affirm Peter, Paul, and other dead men to live in heaven. Now this is the foundation not only of Purgatory but also of that

[71] "Paradoxical Questions Concerning the Morals & Actions of Athanasius & his Followers," William Andrews Clark Memorial Library, MS. shelfmark N563M3 P222, https://www.newtonproject.ox.ac.uk/view/texts/diplomatic/THEM00010," fol. 54[r]. This passage is cited in Snobelen, "Isaac Newton, Socinianism, and 'the One Supreme God,'" 263.

[72] "Paradoxical Questions," fol. 55[r].

horrible idolatry practiced among the Papists, whilst they invocated the saints that are dead."[73]

Newton denied as well disembodied evil spirits. If Milton's sense of divine justice dictated that sinning souls could not survive while sinning bodies died, Newton's emphasis on divine omnipotence left no space for rivals to God, including Satan and demons. Snobelen has argued that "Newton's powerful monotheism, sharpened by his antitrinitarian view that centralized the powers of Deity in the Father alone and combined with his profound, over-arching sense of God's absolute dominion and unchallenged sovereignty, probably also played an important role in Newton's banishment of demons and that chief opponent of the Deity, Satan himself."[74] Newton dismissed stories such as St. Anthony's temptation by demons, a favorite subject of medieval and early modern painters, as monkish impostures.[75] For Newton, the evil spirits of the Bible are allegorical figures for evil inclinations and distempered minds:

> The spirits of God of fals Prophets & of Antichrist are here plainly taken not for any substantial spirits but for the good or evil dispositions & true or fals perswasions of mens minds.... [T]he old Serpent was no more a real serpent then the Dragon in the Apocalyps is a real Dragon or then the Beasts in Iohn & Daniel are real Beasts. Tis only a symbol of the spirit of delusion & therefore the curs of this serpent for deceiving Eve must be interpreted accordingly.[76]

For Newton demons, like the separated spirits of the dead, are not verified by the Bible. If Milton's concern is for the justice of the application of the punishment of death, Newton is wary of any potential avenue for idolatry, for prayer to or fear of any spiritual agency but God the Father.

While an array of Christian theologians across the denominational spectrum argued for the compatibility of Scripture and reason, Socinians were notable for the strength of their claims for reason. In interpreting Scripture, Socinians emphasized like many other Protestants the sufficiency and clarity of Scripture, but they placed a higher premium on the availability of important Scriptural teachings to reason and the apprehensibility of those truths to the sense of the common reader. Appeals to mystery were viewed skeptically. In a brief tract on *Rational Religion*, the Socinian Andrew Wiszowaty writes,

[73] Stegmann, "Brief Inquiry," cited from Williams, *Polish Brethren*, 2:363.

[74] Stephen D. Snobelen, "Lust, Pride, and Ambition: Isaac Newton and the Devil," in *Newton and Newtonianism: New Studies*, ed. James E. Force and Sarah Hutton (Dordrecht: Kluwer, 2004), 171–72.

[75] Snobelen, "Lust, Pride, and Ambition," 166.

[76] "Untitled Treatise on Revelation," Yahuda MS. 1, National Library of Israel, Jerusalem. https://www.newtonproject.ox.ac.uk/catalogue/record/THEM00044," 20^{v}.

> By using...sane reason, man ought to judge by the norm and rule of the written word of God..., affirmed and rightly applied, concerning controversies set forth about religion, what in fact of the truth inheres in them, and to reject those things that are shown to be demonstrably false and evidently absurd....
>
> Once, the judgment of reason is rejected, what will prevent one from seizing upon whatever words of Sacred Letters with a blind drive to deduce the most absurd conclusions whatsoever as if following from them.[77]

One of those absurd conclusions was trinitarianism. Milton for his part stresses the availability of Scriptural truths to reason when arguing against the idea multiple persons in God:

> For it was right, and in highest accord with reason, that that first—and indeed greatest—command, in which God expressed his wish to be dutifully worshipped by a whole people, even the lowest, should be issued in such a way that nothing ambiguous in it, nothing obscure, might drive his worshippers into error, or else keep them in suspense through some doubt. (*DDC* I.2; *OCW* 8.1:41)

Newton likewise emphasized the availability of the divine truths in Scripture as read with the aid of reason; those necessary for salvation, the "milk for babes," are particularly plain and perspicuous.

A confidence in reason as a means to understand the Bible is not the same thing as confidence in philosophy as a guide. Both Milton and Newton echo the Socinian attack on the pernicious importation of Greek philosophy and philosophical terms into Scriptural interpretation. The Socinian Paul Best writes in his 1647 *Mysteries discovered*, "for hypostaticall union and communion of properties, they are but reall contradictions, and the froglike croaking of the Dragon, the beast and false Prophet, *Revel.* 16. 13. by vertue of a *Hocus Pocus* and a Babylonian mouth, thus after the precipice of this Romish *Jezabel*, and the death of her two daughters, *Homousia* and *Symousia* like *Aholah*, and *Aholibah, Ezech.* 23."[78] Best here relates the familiar Greek terms *Houmousia* and the less familiar Greek term *Symousia*, "same essence" and "together in essence," to the daughters by whom Ezekiel allegorizes Samaria and Jerusalem as whoring with idolatrous Egyptians, Assyrians, and Babylonians, associating them through the "Romish *Jezabel*" and "*Hocus Pocus*" (a pointed, intentional corruption of the ritual words

[77] Andrew Wiszowaty, *Rational Religion, or a Tract Concering the Judgment of Reason to Be Used Even in Theological and Religious Controversies* (1685), reprinted in *The Polish Brethren*, 2:677, 682.

[78] Paul Best, *Mysteries Discovered, or, A Mercuriall Picture Pointing Out the Way from Babylon to the Holy City for the Good of All Such as During that Night of Generall Errour and Apostasie, 2 Thes. 2.3. Revel. 3.10 have been so Long Misted with Romes Hobgoblin*, By me PAUL BEST Prisoner in the Gatehouse, *Westminster* (London, 1647), 14–15; Early English Books Online, ProQuest doc. 2264183500, accessed April 1, 2018.

of transubstantiation, *hoc est enim corpus meum*) with the Whore of Babylon, the Roman church.

While Milton, as we have seen, accepts the hypostatical union as a doctrine that, while defying reason, has clear Scriptural warrant, he does, like Newton and Socinians, part company with Arius on the importation of the Greek philosophical such as *homoosios*/homoosios into Christology. In *Of True Religion*, as we saw above, Milton attacks the employment of misleading and specifically trinitarian philosophical terms: "as for the terms of Trinity, Triniunity, Coessentiality, Tripersonality, and the like, they [Arians and Socinians] reject them as Scholastic Notions, not to be found in Scripture, which by a general Protestant Maxim is plain and perspicuous, abundantly to explain its own meaning in the properest words" (*CPW* 8:424–25). Milton fully endorses the plainness and perspicuity of Scripture in matters essential for salvation. This position that he ascribes to Arians and Socinians here he champions himself in *De Doctrina Christiana*. Trinitarians, he charges, write "with the object, not of being seen to proclaim the very simple truth of the gospel to the simple and the poor but, by adopting from amid the barbarity of the schools certain amazing false props of words and sophistries, of supporting with the utmost exertion and persistence some quite ridiculous paradox" (*DDC* I.5; *OCW* 8.1.145). Of this effort, he concludes, "let it completely collapse!" Milton, like Best and other Socinians, explicitly attacks Greek philosophical terms, challenging those who deny that the "one God" of the Bible is the Father alone and who assert trinitarian conceptions of three persons in one essence:

> They beg two questions, as the stock phrase goes, or rather they try to wheedle out of us two free gifts for themselves: first, that the name of God attributed throughout to the Father alone be understood *ousiōdōs*, not *hypostatikōs*, that is, that the name of one Father be taken to mean three persons, or the whole essence of the Trinity, not the one person of the Father. This distinction is not only in many ways ridiculous; it is also thought up merely in order to bolster their opinion, although in reality it does not bolster it but is bolstered by it; and likewise if you invalidate the opinion itself—that is, simply deny it—that empty distinction falls with it. And it is not just an empty distinction, but none at all; more than that, it is 'the same proved by the same', cunningly galvanized by Greek adverbs in order to dazzle the eyes of novices. (*DDC* I.5; *OCW* 8.1:151)

For Milton as for Socinians and Newton, scriptural language of the "one God" is perspicuous, and the importation of irrelevant and unwarranted Greek philosophical terms is a strategy by idolaters to obfuscate the clarity of Scripture.

Newton, while professing an Arian understanding of the relation of Father as eternal, unbegotten begetter of the non-eternal and non-coessential Son, and while defending Arius against the attacks of Athanasius, faulted Arius for his use of non-biblical Greek philosophical terms:

> In these disputes Arius & Athanasius had both of them perplexed the Church with metaphysical opinions & expressed their opinions in novel language not warranted by scripture. The Greeks to preserve the Church from these innovations & metaphysical perplexitys & put an end to the troubles occasioned by them anathematized the novel language of Arius in several of their Councils, & so soon as they were able repealed the novel language of the homousians, & contended that the language of the scripture was to be adhered unto.[79]

A bitter enemy of Athanasius and the Athanasian Creed and a defender of Arius against Athanasius, Newton nevertheless faults Arius for falling into the trap of sophisticating and thus falsifying Scriptural teaching by contaminating it with irrelevant and misleading Greek philosophical terms.

The library of Trinity College, Cambridge, held two copies of the Socinian Best's *Mysteries discovered* during Newton's time as a student and Fellow there. Had he consulted the book, he would also have found language placing the remnant of the faithful against the persecuting idolatrous majority, language foreshadowing his own and resembling Milton's. Best writes, "there is little hope either for the liberty of the Subject, or Law of God amongst us, *Psal.* 119. 126. so this wo will not depart untill it rest in a poor and terrified remnant, as *Revel.* 11. 13. And I cannot understand what detriment could redound either to Church or Common wealth by the toleration of religious [religions?], not antipoliticall, but rather benefit, as we see by example in *Holland* and *Poland*."[80] (With "*Holland* and *Poland*," Best refers to the Polish Brethren, or Socinians, who fled to Holland both before and at the time of their decreed expulsion from Poland in 1660.) Newton, as we have seen, also counted himself a member of a vanishingly small community of prophets, which he described as "a remnant, a few scattered persons which God hath chosen,…set themselves sincerely & earnestly to search after truth."[81]

Adherence to the Apostles' Creed, as opposed to the trinitarian and thus idolatrous Nicene and Athanasian Creeds, served as a sign for Socinians of alignment with the remnant of true interpreters and prophets. The Socinian Jonas Szlichtyng's work on the proper Christian confession follows clause by clause the Apostles' Creed, which lacks the Nicene Creed's affirmation of the Son's being "true God from true God, begotten not made, one in Being with the Father" and the Holy Spirit's "proceed[ing] from the Father and the Son" and who is to be "with the Father and the Son…worshipped and glorified." Szlichtyng begins:

[79] Newton, "Drafts on the History of the Church," fol. 154^{r}.
[80] Best, *Mysteries Discovered*, 14. [81] "Untitled Treatise on Revelation," 1^{r}.

> There is no doubt by anyone that the Creed which is called Apostolic is very ancient, and contains the simplest Apostolic truth itself, such as was handed down from the beginning. For us, this was the cause, that when we wished that it should be testified to the whole world that we pursue primeval truth itself, in bringing out this confession of our churches we followed closely upon the footsteps of this Creed.[82]

Milton, coming to the close of his chapter on the Son of God, by far the longest in *De Doctrina Christiana*, repeatedly endorses the Apostles Creed. Noting that Luke 1:32 describes Jesus as the "*son of the Highest*," Milton adds, "not therefore himself the highest. The apostles universally confirm the same point" (*DDC* I.5; *OCW* 8.1.225). Milton begins and ends the final paragraph of this crucial and signally heterodox chapter with endorsements of the Apostles Creed and thus a repudiation of the offending trinitarian claims of the Nicene Creed: "This was the faith of the saints regarding God's son; this was also the renowned confession of that faith; this alone is taught, is acceptable to God, and holds promise of eternal salvation.... Lastly, the Apostles' Creed...—the most ancient and most accepted creed in the church [*in ecclesia antiquissimum et receptissimum*]—sets before us no other faith" (*DDC* I.5; *OCW* 8.1:227–29).

Newton for his part is exhaustive in his condemnation and refutation of the Nicene Creed; in the massive manuscript "Of the Church" he meticulously untangles the late historical addition of each of the assertions the Nicene Creed adds to the Apostles' Creed. He argues, for example, that the article on "the consubstantiality of the Father & Son" was strong-armed into the Creed at the Nicene Council by the Emperor Constantine in support of Alexander, Bishop of Alexandria, and in opposition to Arius, "& so the Council in the presence of the Emperor after some disputing agreed to this Creed. We beleive in one God, the Father Almighty, the maker of all things visible & invisible, & in one Lord Jesus Christ the Son of God, the only begotten of the Father, that is, of the substance of the Father, ~~by whom all things were made~~ God of God, light of light, very God of very God...."[83] Newton traces the late additions of other articles, including the

[82] Jonas Szlichtyng, "Confession of Christian Faith Published in the Name of the Polish Churches Which Profess One God and His Only-Begotten Son Jesus Christ with the Holy Spirit" (Wrocmirowa, 1642), reprinted in in *The Polish Brethren*, 2.389–414 (passage quoted from 389).

[83] Newton, "Of the Church," (Fondation Martin Bodmer, Geneva, Switzerland), Chapter 3, fols. 9^{r}–10^{r}. Newton in the omitted section details the machinations, citing in the margin his sources (Socrates Scholasticus' *Ecclesiastical History* I.8 and Sozomon I.12): "When the Emperor Constantine the great had convened the Council of Nice, & the Bishops could not agree, some of them being for Alexander Bishop of Alexandria, some perhaps for Arius & some for silencing the dispute without ~~setting up either party~~ admitting the new language of either party, the Emperor upon a day appointed came into the Council to bring things to a conclusion: & Eusebius of Cæsarea proposing the Creed above written the Emperor & all the Council approved thereof. But the Emperor being for Alexander, proposed & pressed that the consubstantiality of the son should be inserted, & so the Council in the presence of the Emperor after some disputing agreed to this Creed."

Son of God's possessing a human as well as a divine nature (15^{r}), his descent into Hell (15^{r}), the worship of the Holy Ghost (16^{r}), the resurrection of the body (17^{r}), the Son's sitting at the right hand of the Father (20^{r}). Newton argues that "if the new articles be omitted, the Creed of the Latins called the Apostles Creed will be reduced to this form."[84]

Believers' baptism is a key Socinian imperative not associated with Arianism but shared by various dissenting sects. For Socinians baptism is not a ritual with power to remove sin but an external sign of a believer's inner assent to Gospel teaching. The *Racovian Catechism* puts it this way:

> Q. *What think you concerning the Baptism of Water?*
>
> A. That it is an externall Rite, whereby men coming from Judaisme, or Gentilisme to the Christian Religion, did professe openly, that they acknowledge Christ for their Lord.
>
> Q. *Do Infants belong to that Rite?*
>
> A. By no means, for neither have we in the Scripture either precept, or example thereof; nor can they, as the thing it self sheweth, acknowledge Christ for their Lord.
>
> ....
>
> Q. *What think you of them that think they are regenerated by this Rite?*
>
> A. They are exceedingly mistaken, for Regeration [sic] is nothing but the transformation of our mind, and will, and composure of them to the doctrine of our Saviour Christ, as the very word [Regeneration] doth intimate. But such a transformation cannot have place in Infants, who know not good and evill, much lesse that a thing of so great moment should be incident to them.[85]

The Summary of the Socinian Convocation at Roznów likewise points to the inability of infants to give the assent required of the baptized: "All Christians baptize infants, and only the Brethren [i.e., the Polish Brethren, or Socinians] fail to do so, since they emphasize that rational faith must precede the rite."[86]

Milton likewise views baptism as a voluntary sign of acceptance of the Word, a process of which infants are as yet incapable:

[84] "Of the Church," Chapter 2, fol. 40^{r}. What follows the statement in the manuscript is the Apostles' Creed: "*I believe in God, the Father almighty, maker of heaven & earth: & in Jesus Christ his only son our Lord, who was born of the Virgin Mary, suffered under Pontius Pilate, was crucified, dead & buried, the third day he rose again from the dead, & ascended into heaven & sitteth at the right hand of God, & shall come to judge the quick & the dead. And I beleive in the holy Ghost.*"

[85] *Racovian Catechism*, 109–10. See also Peter Morzkowski's *The Ecclesiastical Polity Which They Commonly Call the Agenda or The Form of the Exterior Government of the Chritian Churches of Poland, Which Confess One God the Father through His Only Begotten Jesus Christ in the Holy Spirit* (1646), Item C2, reprinted in Williams, *Polish Brethren*, 2:448–49.

[86] Summary of "Final Session, 16 March 1660, on Pedobaptism," reprinted in Williams, *Polish Brethren*, 2:624.

> infants are not capable of baptism, seeing that they can neither be taught nor believe nor make a pledge, nor give either an assurance or an answer, [and] cannot even hear the word. For how are infants, who cannot understand the word, purified through the word?... [B]oth intellect and will are required for baptism, neither of which infants possess. Hence that statement in Acts 8: 37: *if you believe from within your whole heart, it is permitted to be baptized.* But surely infants cannot believe from within even the smallest particle of their heart. (*DDC* I.28; *OCW* 8.2:733, 737)

With characteristically caustic wit, Milton asks of believers in infant baptism, i.e., Catholics and the majority of Protestants, "Who would not be ashamed of such infantile interpreters?/*Quem non pudeat interpretum tam infantium*?" (*OCW* 8.2:737, 736).

Newton also insists that belief in the original faith is required for baptism, and thus that infant baptism is illegitimately performed. In his draft chapter "Of the holy covenant" in "Of the Church," Newton insists that authentic baptism requires believers to profess faith in that Creed before being baptized: "Thus Christ after his resurrection instructed his Apostles in the faith & practise of the Christian religion & commanded them to teach all nations the same faith & then to baptise them teaching them also repentance & a good life.... It was the faith rule & {illeg.} *faith wch all men were to learn before baptism* & wch was sufficient for baptism."[87] Newton elaborates on the requirements for baptism in the "Irenicum." After quoting the Apostles' Creed, he writes, "All this was taught from the beginning of the Gospel in Catechising, that the Catechumen might know before Baptism.... ffor every thing necessary to communion must be taught before admission into it."[88]

The *Racovian Catechism* counsels peaceful coexistence with those who believe, erroneously, in infant baptism: "Although they erre herein, yet is it not therefore lawfull to condemn them, so that they be not otherwise Idolaters, but live piously according to Commandments of Jesus Christ, and forbear to persecute others, who reject their opinion."[89] The great emphasis by Socinians and other dissenting sects on toleration is natural given their acute awareness of their minority positions on such fraught issues as the trinity, the immortality of the soul, and infant baptism, and it is shared for similar reasons by Milton and Newton.[90]

[87] "Of the Church," Chapter 2, fol. 26^{r} (my emphasis).

[88] "Irenicum, or Ecclesiastical Polyty tending to Peace," Keynes Ms. 3, King's College, Cambridge," fol. 1;

[89] *Racovian Catechism*, 109.

[90] For Socinian arguments for toleration, see the following works, reprinted in *The Polish Brethren*, and cited by Snobelen, "Isaac Newton, Heretic," 387: Samuel Przypkowski, *A Discourse Touching the Peace and Concord of the Church* (1628, 1630; trans. John Dury 1688), 1:291–302; Johannes Crellius, *A Vindication of Liberty of Religion* (1637; trans. John Dury, 1646), 1:342–54; and Stanislas

In distinguishing innocent doctrinal error from willed contradiction of Scriptural truth, as outlined in the previous chapter, Milton and Newton Milton resembled Socinians. Blameable heresy for them and professed Socinians lay in the corrupt will of the interpreter, not in the honest error of the sincere reader. Milton asks in *Of True Religion* how one might answer the Papist who asks if there are no heretics among the various sects of magisterial and radical reformers:

> I answer, all these may have some errors, but are no Hereticks. *Heresie is in the Will and choice profestly against Scripture; error is against the Will, in misunderstanding the Scripture after all sincere endeavours to understand it rightly* [my emphasis]. Hence it was well said by one of the Ancients, *Err I may, but a Heretick I will not be.* It is a humane frailty to err, and no man is infallible here on earth. (*CPW* 8:423)

The Socinian Samuel Przypkowski makes the same argument in *A Discourse Touching the Peace and Concord of the Church*, translated by John Dury, Milton's acquaintance and translator of *Eikonoklastes* into French. Przypkowski heads a chapter with the claim that "nothing but disobedience and unbelief exclude a man from eternal salvation; and that such as err are free from these."[91] He elaborates,

> It now remaineth to be considered whether bare errors of the mind are disobedience. That they are not so is even from hence apparent, because they consist in the mistaking of the understanding, but this in the depravity of the will. Again, since God regardeth not the actions themselves but the mind of the agents (which the Greeks call proaivresiß), who seeth not, that they who, desiring from their hearts to obey, yet through error or imprudence do not see, cannot be justly accused of disobedience.[92]

Those who err should be tolerated, unless their intention is to ignore or corrupt Scriptural teaching rather than to sincerely search and teach it.

Milton and Newton agreed with Socinians that state coercion should have no place in religious controversy. Without the free expression of religious differences, Milton writes, "there is no religion, no Gospel; violence alone prevails; but it is disgraceful and shameful that the Christian religion stand upon violence: our slavery endures still; not as previously to divine law, but what is most wretched of all, to human law; or, so to speak more accurately, we must be slaves to *in*human

Lubieniecki, *The Legal Claims for the Freedom of the Religion of Unitarians in Poland* (1658 or 1660 or 1662), 2:559–81.

[91] Przypkowski, *A Discourse Touching the Peace and Concord of the Church*, in *The Polish Brethren*, 1.299.

[92] Przypkowski, 1.299.

tyranny" (*DDC* Epistle; *OCW* 8.1:9). Milton here echoes the the argument of the leading Socinian intellectual of the seventeenth century, Johannes Crellius, whose *Vindication of Liberty of Religion*, was also translated by John Dury:

> if Catholics should deny Heretics liberty or peace and should hold it lawful to punish them and so should punish them for professing that religion which differs from the Catholic, they would force them, as much as in them lay to their own religion. But no man ought to be forced or compelled by outward violence to the Christian religion. First, because the disposition and nature of religion, especially spiritual, and perfect religion, such as is the Christian religion, is repugnant to compulsion.[93]

Crellius argues in a Continental context where there is real danger of persecution by the Roman Catholic church. Reflecting a different context, Milton, unlike Crellius, opposed toleration of Roman Catholicism, on the grounds that it would not itself practice the kind of toleration for which Crellius appeals.

Newton also called for toleration. He addressed doctrinal "strong meats," or things not necessary to salvation:

> About these & such like things we are ^to believe the scriptures,^ to learn all that we can ^~~by studying the~~^, & to instruct one another ^& therefore not to impose consequences of scripture but to keep to ye language of the scripture it selfe^ but not to fall out about them. The strong in faith & knowledge must not despise the weak, & the weak must not judge & censure the strong, but all Christians keeping the common faith into which they were baptized & whereby they wer admitted into the Church & united into one mystical body of Christ must bear with one anothers infirmities.[94]

Newton put into practice this conviction, serving on committees for the relief of French Protestant ministers and for the toleration of dissenters."[95]

With Socinians, Milton and Newton argued for the separation of church and state and opposed religious coercion by the civil arm.[96] Milton elaborates on the arguments for toleration in his 1659 *Treatise of Civil Power in Ecclesiastical Causes*, where he asserts the antithetical nature of violence or coercion and religious profession and insists that civil authorities have no right to compel religious belief and practice: "No protestant... of what sect soever following scripture only,

[93] Crellius, *Vindication of Liberty of Religion*, Chapter 3, "For Freedom of Religion," excerpted in *The Polish Brethren*, I.342–55. The quoted passage is on 344.

[94] Newton, "Of the Church," Chapter 3, fol. 24^{r}.

[95] See Scott Mandelbrote, "'A Duty of the Greatest Moment': Isaac Newton and the Writing of Biblical Criticism," *British Journal for the History of Science* 26 (1993): 288.

[96] For Socinian support for the separation of church and state, see George H. Williams, *The Radical Reformation*, 3rd ed. (Kirksville: Sixteenth Century Journal Publishers), 1282–84, cited in Snobelen, "Isaac Newton, Socinianism, and 'the One Supreme God,'" 264.

which is the common sect wherin they all agree, and the granted rule of everie mans conscience to himself, ought, by the common doctrine of protestants, to be forc'd or molested for religion" (*CPW* 7:254). Newton argues likewise against state power over religious conscience in the manuscript "Notes on early church history and the moral superiority of barbarians to the Romans":

> The Magistrate may punish or cut off any for their vices or evil actions but not professors of Christianity for erronious opinions, least they pluck up the Wheat with the Tares. The Church may reprove or excommunicate but she has as little authority to guide the arm of the Magistrate as to handle his sword: for this is to make her self the judge & him but the executioner. Shee may excommunicate but not force into communion. Christ never instituted that a means of her propagation & preservation. If we would have them once with us we must use the proper means to beget faith in them, & not urge them by violence to do what is contrary to their perswasion, seing whatsoever is not of faith is sin.[97]

Newton's warning here against "pluck[ing] up ye wheat wth ye Tares" echoes Milton's argument in *Areopagitica* that those who would suppress religious dissent assume erroneously that they are in possession of the truth, which is a "streaming fountain... flow[ing]... in a perpetuall progression," when in fact "It is *not possible for man to sever the wheat from the tares*, the good fish from the other frie; that must be the Angels Ministry at the end of mortall things" (*CPW* 2:543,564–65; my emphasis).[98]

Differences with Socinians

Neither Milton nor Newton, as I noted at the beginning of this chapter, was a Socinian. Their disagreements with key elements of the Socinian program—on free will, on original sin, on the nature of the Son, and the Atonement—once again highlight their differing theological imperatives, universal freedom and divine omnipotence respectively.

A particularly significant area of disagreement among Socinians, Milton, and Newton concerns the temporal origin and the nature of the Son. Earlier in this

[97] Yahuda Ms. 39, 1^{r-v}; https://www.newtonproject.ox.ac.uk/view/texts/normalized/THEM00076, accessed January 29, 2024.

[98] Milton alludes, as does Newton, to the parable of the wheat and tares (Matt 13:24–30, 36–43). Milton returns to the parable at the end of the chapter "On the Gospel and Christian Freedom" in *De Doctrina Christiana*. For the magistrate to compel religious uniformity and remove from truth-seekers the freedom of interpretation, "he removes the **Gospel** itself; at the least, he removes the good and the evil people together, contrary to that very well-known command, Matt. 13: 29–30: [*no*, i.e. do not gather up the tares,] *lest by gathering up the tares you uproot the wheat at the same time with them: let both grow until the harvest* (*OCW* 8.2:723; bracketed verse in the Oxford translation).

chapter I contrasted Milton's reluctant acceptance of the hypostatic union of divine and human natures in the Son with Newton's denial of that union and assertion that the Son does not have a human nature. There, I suggested that Newton's position accorded with his insistence on an infinite distance between the omnipotent God and created human beings, while Milton's corresponded with his tendency to compress ontological hierarchies. Milton's Son shares more with us than Newton's. The Socinians position on the hypostatic union is the remaining alternative. Their Son has neither a dual nature, as does Milton's, nor a single divine nature, as does Newton's. Socinians believed that the Son has one, human nature, and that he came into existence when he was conceived in Mary by God. Milton, in the chapter on the Son of God in the *De Doctrina Christiana*, repudiates the Socinian teaching on when the Son came into being: "This is certain—whatever certain revisionists prattle to the contrary—that the Son, under the name of the word or Speech, existed in the beginning and was the first of created things, through whom all the rest were then made" (*DDC* I.5; *OCW* 8.1:129). The Socinian Son is unique as the only human being born of a woman but fathered by the Father, and he is worthy of secondary worship because of his role in redemption, but his nature is not divine. The *Racovian Catechism* lays out the logic for rejecting the divine nature of the Son:

> Q. You said a little before that the Lord Jesus is a man by nature, hath he not also a divine Nature?
>
> A. At no hand; for that is repugnant not onely to sound Reason, but also to the holy Scriptures.
>
> Q. Shew me how it is repugnant to sound Reason.
>
> A. First, because two substances indued with opposite properties, cannot combine into one Person, and such properties are mortality and immortality; to have beginning, and to be without beginning; to be mutable, and immutable. Again, two Natures, each whereof is apt to constitute a severall person, cannot be huddled into one Person. For instead of one, there must of necessity arise two persons, & consequently become two Christs, whom all men without controversie acknowledge to be one, and his Person one.[99]

The logic here resembles that which informs Milton's and Newton's Arian positions. That which has a beginning and that which, *a fortiori*, is mutable, cannot be the supreme, eternal, immutable God. But where Milton and Newton conclude that the Son is endowed by the Father with delegated and subordinate divinity, Socinians reject this option. One is either not divine or fully divine, with no possibility of a delegated or subordinate divine nature. They reject the

[99] *Racovian Catechism*, 28.

"huddl[ing]" of "two Natures . . . into one Person," which was accepted by Milton because he finds Scriptural warrant despite the its affront to reason ("*how* it is so, is not known, and it is assuredly best that what God wills to be unknown should not be known" [*DDC* I.14; *OCW* 8.1:481]).

The passage from the *Racovian Catechism continues,*

> Q. But when they alledge that Christ is so constituted of a divine and humane Nature, as a man is of a body and soul, what answer must we make to them?
>
> A. That in this case there is a wide difference; for they say that the two Natures in Christ are so united, that Christ is both God and Man. Whereas the soul and body in a man are so conjoyned, as that a man is neither soul nor body. For neither doth the soul nor the body severally constitute a Person. But as the divine Nature doth by it self constitute a Person, so must the humane by it self of necessity also constitute.
>
> Q. Shew how it is also repugnant to the Scripture that Christ should have a divine Nature.
>
> A. First, because the Scripture proposeth to us but one God by nature, whom we formerly demonstrated to be the Father of Christ. Secondly, the same Scripture witnesseth that Jesus Christ is a man by nature, as was formerly shown.[100]

Again the logic resembles Milton's and Newton's. Neither the soul nor the body constitutes separately a person. The separated soul, if such a thing were possible, would not be a person. For Milton, Newton, and the Socinians, the person dies entirely and is reconstituted at the resurrection of the dead as once again a person with integrated and inseparable body and soul.

Newton shares the Socinian objection to bundling two natures into one person, but as we have seen he rejects the ascription of a human nature to the Son as, like the Trinity itself, a late innovation. The early church, according to Newton, viewed the Son's nature as divine (if subordinately and non-coessentially so):

> Another new Article inserted into the Creed was that the Son of God assumed a man consisting of a body & soul & united this man to himself, & by this union the two natures became one Christ. This — Article was ~~first~~ proposed ^first^ . . . by Athanasius & some other Bishops convened in a Council at Alexandria in the [same] reign of Jovian, & afterwards established by the third & fourth General Councils & at length made an article of the Creed commonly ascribed to Athanasius.[101]

That Newton believed the idea of hypostatic union of two natures in one Christ originated in Athanasius, a figure he despised as a mendacious and murderous

[100] *Racovian Catechism*, 28–29.

[101] Newton, "Of the Church," Chapter 3, fol. 15r.

Machiavellian *avant la lettre*, might be enough to turn him against the doctrine. But there is another and more compelling reason. For Newton, the Son is the obedient instrument of an omnipotent God the Father, mediating divine commands in creating, in establishing the laws governing creation, and in dealings with rational creatures. While Newton believes that the non-eternal begotten Son of God did become incarnate, because God the Father lies at an infinite distance from his creation, it is fitting that the mediating Son, uniquely begotten by the Father and through whom he created all else that is, lies at a great distance from that creation, including rational creatures. Jesus Christ is here for Newton not the union of God's Son begotten before creation with the human son of Mary, but a merely divine but subordinate figure.

Nevertheless, Newton, despite his at times adamant support of the Arian position that the Son of God, while not eternal, preexists Creation, surprisingly sometimes treats a crucial disagreement between Arian and Socinian Christology—between a preexistent Son of God through whom God creates everything else and a Son of God who came into existence when conceived in Mary—as one of things indifferent that should not cause conflict among Christians. In his emphatic endorsement of the Apostles' Creed over the Nicene Creed in the manuscript draft chapters of "Of the Church," Newton writes that the preexistence of Christ, the Arian position he favors, is a late addition, a circumstance that normally leads him to question a doctrine:

> The Article, *Begotten before all ages*, was not in the Creeds of Rome, Aquileia, Turn, Ravenna, nor in those of Afric. I meet with it in none of the Creeds of the Latin Churches nor in those of Tertullian & Justin Martyr. For those who beleived that Jesus Christ took his beginning of the Virgin continued in communion wth the Churches in his [Justin Martyr's, 100-c165 CE] days & even till the days of Pope Victor [c189–199 CE].[102]

Newton then observes that it was not until later, in the third century, that the adoptionist argument—that the Son came into existence when born of Mary—was condemned. He adds that the article "begotten before all worlds" began to be inserted around the Council of Nicaea. Here and elsewhere in "Of the Church," Newton points to the fact that Arian and adoptionist, or what would come to be called Socinian, opinions on the temporal origin of the Son co-existed peacefully without schism or charges of heresy.[103]

Milton and Newton also departed in different ways from Socinian teachings on free will and original sin. Socinians were strong and unambiguous advocates

[102] Newton, "Of the Church," Chapter 2, fol. 34^{v}.

[103] See Snobelen, "Isaac Newton, Socinianism, and 'the One Supreme God,'" 251, for a key passage from "Of the Church," Chapter 5a, fol. 3^{r}.

of the freedom of the will, and unlike Milton they flatly denied the effects of original sin on the will. Sarah Mortimer describes the free will argument, and the legalist terms, of *De Vera Religione*, a work begun by one important Socinian, Jan Volkelius, and finished by another, Johan Crell:

> the authors were insistent throughout that God had given free will to human beings and that any covenant between God and men must be conditional. Crell...assumed that God was a sovereign ruler, but one whose relationship to human beings could be understood in terms of rights. On Crell's account, God gave to men free will and thereby made them at least in part *sui iuris*, able to act in a legal capacity in their own right, even in matters pertaining to their salvation.[104]

In this emphasis on free will, Crell follows the *Racovian Catechism*'s tenth chapter, "Of Free-will." The ninth chapter ends with a definition of the obedience God requires "of us, who believe in Christ," which specifies that Christian obedience involves "contract[ing] the habit of no sin, but gain[ing] to our selves the habits of all Christian Vertues." This "habit of no sin" seems to conflict with orthodox claims of the limits of fallen nature, and thus the tenth chapter begins thus:

> Q. *Doth it lie in our power to obey God in that manner?*
> A. Yes, for it is certain, that the first Man was so created by God, as that he was endued with Free-will. Neither was there any cause why God, after his fall, should deprive him thereof. And indeed the Justice of God doth not permit it, nor is there any mention of such a punishment, amongst those punishments that God inflicted on *Adam* for his transgression.[105]

Milton for his part repeatedly affirms the freedom of the will, in *Areopagitica*, *De Doctrina Christiana*, and *Paradise Lost*. *Areopagitica* affirms the freedom of the prelapsarian Adam and Eve in the strongest terms: "when God gave [Adam] reason, he gave him freedom to choose, for reason is but choosing; he had bin else a meer artificiall *Adam*, such an *Adam* as he is in the motions" (*CPW* 2:527). Milton's God expatiates at length on the freedom of the first couple:

> I made him just and right,
> Sufficient to have stood, though free to fall.
> Such I created all th' ethereal Powers
> And spirits, both them who stood and them who failed;
> Freely they stood who stood, and fell who fell.
> Not free, what proof could they have giv'n sincere

[104] Sarah Mortimer, *Reason and Religion in the English Revolution: The Challenge of Socinianism* (Cambridge: Cambridge University Press, 2010), 31.
[105] *Racovian Catechism*, 141–42.

Of true allegiance, constant faith or love?
Where only what they needs must do, appeared,
Not what they would? (*Paradise Lost* 3.98–106)

Two lines later Milton's God echoes ("reason also is choice") the author of the *Areopagitica* ("reason is but choosing").

In the same speech, however, the Father proclaims "I formed them free, and free they must remain, / Till they enthrall themselves" (*PL* 3.124–25). The language evokes the "bondage of the will," a central theme of Paul, Augustine, and Luther, who insist that original sin radically curtails our freedom. The reference to self-enthrallment might suggest that Milton rejects Socinian skepticism concerning original sin. The *Racovian Catechism* continues where the earlier passage leaves off:

Q. *Is not this Free-wil depraved by Originall sin?*

A. There is no such thing as Originall sinne, wherefore neither can it deprave Free-will. For that Originall sinne cannot be proved out of the Scripture. And the fall of *Adam* being but one act, could not have such power as to deprave his own nature, much lesse the nature of all his posterity. But that this punishment was inflicted upon *Adam*, neither doth the Scripture any where relate, and it is altogether unsuitable to God, who is the fountain of all equity, so to do.

Q. *But yet that there is Originall sin, seemeth to be proved by that testimony,* Gen. 6. 5. *All the thoughts of mans heart were only evill, and that continually; and* Gen. 8. 21. *The thought of mans heart is evill from his child-hood?*

A. These testimonies speak of voluntary sin, so quireth of them; But a will to perform them is in all men by nature: yet is not that ability so little, that, if a man will offer violence to himself, he may not by the accession of the Divine Assistance, obey the Divine Will. But [that] God denieth his assistance to that Originall sin cannot be proved thence.[106]

The Socinian rebels at the Augustinian doctrine of original sin. For Socinians, human beings have the power to resist the tendency to sin, and God is ready to offer any necessary assistance, thus steering clear, like Milton, of Pelagianism. Despite Milton's language of self-enthrallment, and despite the crisis of *Paradise Lost* of the "completing of the mortal sin / Original" (*PL* 9.1003–4), heightened by enjambment, Milton goes some way toward a position close to the *Racovian Catechism* in *De Doctrina Christiana*: "when God decided to restore humankind, he also indubitably decreed (for what is fairer?) to restore lost freedom to the will,

[106] *Racovian Catechism*, 142–43.

in some measure at least" (*DDC* I.4; *OCW* 8.1:95). In practical terms, Milton's Arminian assertion of universally offered sufficient grace amounts to an assertion of free will retained after the Fall.

The *Racovian Catechism* outlines its own version of will-enabling grace. God will be there to help the free human being resist evil urges and freely choose to obey. The language of the *Catechism* is not entirely clear on temporal sequence, but it seems to say that if one resists an inclination to evil—"offer[s] violence to himself"—one will receive grace—an "accession of the Divine Assistance"—and be able to obey. While Milton's God argues in *Paradise Lost* that after the Fall grace must come before the good will (3.185–97), Milton's first illustration of the action of such grace in the poem reverses this order, as Adam and Eve offer their prayer of contrition at the end of Book 10 (10.1086–104), before the poem revises the sequence at the beginning of Book 11 to fit the orthodox ordering of contrition *after* prevenient grace: "Thus they in lowliest plight repentant stood / Praying, for from the mercy-seat above / Prevenient grace descending had removed / The stony from their hearts, and made new flesh / Regenerate grow instead" (11.1–5). Despite this retrospective qualification, in the narrative order penitence precedes grace, resonating both with the *Catechisme* and with Milton's suggestion in *De Doctrina Christiana* that God "restore[s] lost freedom to the will, in some measure at least."

This aspect of Socinian teaching is of far less interest to Newton. His assignment of the question to *adiaphora* pales beside his repeated assertions of the Father's absolute omnipotence, his controlling of all actions and events. Newton jealously protects the freedom of the Father. In a world in which God is the absolute ruler of unchallenged dominion, there is less concern for the niceties of divine justice and less solicitousness for the freedom of human beings.

Newton, on the other hand, apparently doubted one orthodox belief with implications for an understanding of divine justice: the eternity of punishment after death. His protégé William Whiston, who denied the eternity of hell fire, claimed that Newton shared his skepticism. Whiston claims that this position was shared also by his fellow protégé, Samuel Clarke: "And I think I may venture to add, upon the Credit of what I discovered of the opinions of Sir *Isaac Newton* and Dr. *Clarke*, that they were both of the same sentiments. Nay, Dr. *Clarke* thought that 'few or no thinking Men were really of different sentiments in that Matter.'"[107] The denial of the eternity of hell fire is common among Socinians. It is

[107] Whiston, *Historical Memoirs of the Life and Writings of Dr. Samuel Clarke* (London, 1730), 75, as quoted in James E. Force, "God of Abraham and Isaac," 189. Mordechai Feingold, pointing to the fact that Whiston waited until after Newton's death to offer his observations and that Whiston was angry at Newton for abandoning him when he was ejected from Lucasian Chair in Mathematics and expelled from Cambridge University, advises caution in accepting Whiston's testimony. See Mordechai Feingold, "Isaac Newton, Heretic? Some Eighteenth-Century Perceptions," in *Reading Newton in Early Modern Europe*, ed. Elizabethanne Boran and Mordechai Feingold (Leiden: Brill, 2017), 338.

argued at length in the *Epitome of a Colloquium Held at Raków in the Year 1601*.[108] Significantly, the Colloquium's argument against the eternity of infernal punishment is based not only on skepticism about the eternity of hell but also on an assertion of the freedom of God's will and action: "The justice of God requires that he punish no one undeserving. However, it is licit for God to dispense with punishing a malefactor, because he is a supreme judge.... Death itself is a punishment, even if none of the dead feel it, because none will ever be liberated from it, from which, however, one could have been freed, if one had obeyed God."[109] Like the Socinian God, Newton's is not bound to punish eternally with torments; eternal death, or nothingness, a state that Milton's Moloch seems to desire (*PL* 2.82–101) and from which Milton's Belial recoils (*PL* 2.142–51), is punishment enough. Newton elsewhere, in a discussion of prophetic terms in the Bible, argues that "hell" is used as an allegorical term describing historical and political realities on Earth: "ascending towards heaven & descending to the earth \or into it/ is put for rising & falling in honour & power \~~& the lower parts of the earth called Hades or Hell \for/ the lowest & most miserable part of the people~~/. Rising out of the earth or waters, or falling into them, for the rising to any dominion or dignity out of the inferior state of people or falling from the same into that inferior state."[110] This skepticism about a literal Hell is related to Newton's denial of the existence of demons, his understanding of their presence in the Bible as figurative for psychological disorders, and his understanding of Satan not as a personal being but "the spirit of deception."[111] Newton in his developing thinking on the punishments of Hell echoes Socinian teaching on God's dominion and unfettered omnipotence. In his denial of personal devils, he goes beyond Socinians in removing any supernatural opposition, however inferior and ineffectual, to God's dominion.

Milton differs from Socinians both in his insistence on infernal punishment and in his argument that divine justice requires that punishment. In *De Doctrina Christiana*, before citing ample scriptural warrant for the eternal damnation in Hell, he points to a key reason that the dead must be resurrected.[112] If God did not ordain general resurrection of the dead, "the righteous would be the most wretched of all people, the impious the most blessed, because their portion in this life is far richer—which would plainly be unworthy of divine providence and justice" (*DDC* I.33; *OCW* 8.2:879). What a Socinian might praise as God's laudable

[108] The *Epitome* records the proceedings a colloquium attended by central founders of Socinian teaching, including Socinus himself, Valentine Schmalz, Stanislas Lubieniecki, and others. For the Socinian arguments, see Item D, "On the State of the Dead up to the Last Day," in *The Polish Brethren*, 1:112–26.

[109] *The Polish Brethren*, 1:115, 116.

[110] Newton, "Two incomplete treatises on prophecy," Keynes Ms. 5, King's College, Cambridge, UK, fol. 2^{r} (http://www.newtonproject.ox.ac.uk/view/texts/diplomatic/THEM00005, accessed April 2, 2018).

[111] "Of the Church," Chapter 2, fol. 21^{r}.

[112] Among the verses Milton cites for the existence of Hell are Isa 30:33 and 66:24, Dan 12:2, Matt 8:12, Mark 9:43, and Rev 14:11 and 21:8 (*DDC* I.33, *OCW* 8.2:891).

mercy, Milton would condemn as divine injustice. For Milton as for Dante, though Milton does not indicate this in his epic, one's punishment in Hell will be proportionate to one's crimes. Citing Matthew 11:22, he argues in *De Doctrina Christiana* that "The degrees of the punishments vary, however, in proportion to the sins," and he adds that "the hell of the damned is indeed the same as that which had been prepared for the devil and his angels" (*DDC* I.33; *OCW* 8.2:893). That hell, we see in *Paradise Lost*, is a place of sensory as well as psychological pain, with its "lake with liquid fire" (1.229), "burnt soil" (1.562), fiery rivers (2.575–81), and, lest the devils (and eventually damned humans) acclimate, "deep snow and ice," where "the parching air / Burns frore, and cold performs th'effect of fire," and where the damned can be hauled "From beds of raging fire to starve in ice" (2.591, 594–95, 600). In his epic hell Milton paints in detail what in his theological treatise he refers to as a place of "eternal torment,...the punishment of sense," which is paired with "punishment of deprivation," or "the loss of the supreme good,...of divine grace and protection and the beatific vision" (*DDC* I.33; *OCW* 8.2:891). The God of Milton's epic affirms the eternity of these punishments. After "Bad men and angels...sink / Beneath [the Son's] sentence[,] Hell, her numbers full, / Thenceforth shall be forever shut" (3.331–33). This is one reason Milton argues that hell cannot be within the Earth, for otherwise the place of punishment would be annihilated; "if this happened," he notes with grim humor, "it would be a really splendid finish for the damned!" (*DDC* I.33; *OCW* 8.2:893).

If Milton and Newton differed from Socinian teaching on the nature of the Son and on the importance of creaturely free will in relation to salvation and divine justice, they approached, although they do not adopt, a Socinian understanding of the relation of Christ's passion to redemption, though for different reasons. Socinus and his followers rejected the notion that the Father had an obligation to substitute the death of his Son for the death of fallen creatures. Moreover, they placed greater emphasis on Christ's free offer of himself as exemplary self-sacrifice than on his death as cleansing sacrifice.[113] The *Racovian Catechism* lays out the argument that in dying on the cross Christ provided an example for imitation rather than a sacrificial atonement:

Q. *But what reason was there that the Saviour should endure the same afflictions and death with the saved?*

A. There are two reasons thereof, as Christ also saveth the Faithfull in a two-fold way; for first, by his example, he moveth them to persist in the way of Salvation that they have entred into. Next, he standeth by them in every combate of temptations, afflictions, and troubles, and at length delivereth them from Eternall Death.[114]

[113] See Snobelen, "Isaac Newton, Socinianism, and 'the One Supreme God,'" 262–63.
[114] *Racovian Catechism*, 122–23.

Johann Crell separates the expiation of humanity's sins from Christ's passion" in his 1646 *Expiation of a Sinner in a Commentary upon the Epistle to the Hebrews*:

> 1. That Christ our high Priest expiateth our sinnes, by succouring us in our temptations. 2. That the principall function of his Priestly office is performed now in heaven and was not performed at this death, wherein there was only a preparation toward it. 3. that neither the Priestly function of Christ, nor his Expiation of sins thereby procured, consist in this, that Christ should suffer punishment for our sins, seeing that can have no place in heaven.[115]

In the Socinian chronology, the offer of the Son's sacrifice cannot come in Heaven before the Incarnation, as he did not preexist his conception and birth from Mary, while in Milton's Arian chronology, the offer comes before Creation in Heaven.

Milton supports the doctrine of the atonement in *De Doctrina Christiana* I.16, where he writes that "Christ as Theanthropos both in life and in death voluntarily submitted himself to divine justice in order to undergo all things by which our redemption was to be accomplished" (*DDC* I.16; *OCW* 8.1:511). In *Paradise Lost* Milton's God is famously and emphatically intent on the legal requirement of substituted death: "Die he or Justice must; unless for him / Some other able, and as willing, pay / The rigid satisfaction, death for death" (3.210–12).[116] Nevertheless, sandwiched between the iron rigor of "Die he or Justice must" and "rigid satisfaction, death for death" is a crucial Miltonic qualifier; the satisfaction must be made by "some other" not only "able," but also "as willing." And for Milton, more than for Newton and more than is possible within the logic of trinitarianism, the Son is free to will his role or not, a freedom vividly dramatized in the scene in Heaven immediately following the Father's speech. It is this radical freedom, as we have seen, that earns him the praise of being "By merit more than birthright Son God" (3.309). Rogers has argued persuasively that Milton's Son, in a chronologically earlier moment nevertheless resembling the exemplary and salvific offer of the Socinian Son, saves humankind not by his sacrificial death but by his free offer, his oblation. Rogers argues, moreover, that Milton, by placing this offer before creation rather than after his death, draws the Son's act, despite its earlier placement in time, closer to the Socinian Son's "freely willed gesture of priestly offering."[117]

[115] Johann Crell, *The Expiation of a Sinner* (London, 1646), 39; Early English Books Online, ProQuest doc. 2264213504; accessed June 21, 2022. Rogers refers to this passage in "Milton and the Heretical Priesthood," 218, n. 12.

[116] On Milton's disagreement with Socinians rejection of the satisfaction theory of atonement, see Jason Kerr, "Felbinger Not the Author of *De Doctrina Christiana*, Currently Attributed to Milton," *Notes & Queries* 28 (2025): 65–68.

[117] Rogers, "Milton and the Heretical Priesthood," 209.

Milton does, it must be acknowledged, explicitly reject the Socinian view of Christ's passion as exemplary rather than atoning: "Those who maintain that Christ faced death not in our place for the purpose of redemption but only for our good and by way of example are vainly trying to evade [Scriptural] evidence" (*DDC* I.16; *OCW* 8.1.521). But as many have pointed out, Milton does not embrace the passion and atonement in his poetry.[118] His early poem "On the Passion," only a fragment, is lacking the energy and conviction of the "Ode: On the Morning of Christ's Nativity." The crucifixion is dispatched in fewer than ten lines in his survey of biblical history in the final books of *Paradise Lost* (12.411–17). In *Paradise Regained*, Milton narrates not the Passion but the Temptation in the Desert, offering the incarnate Son as an exemple of faithful interpretation of Scripture and of virtuous choice.

This way of thinking about the Son of God bridges an otherwise unbridgeable chasm between the Son and the human being. In the more orthodox satisfaction/punishment theory of atonement, no sacrifice of a mere human would suffice; only by the death of the Son of God can the debt humanity incurs by its sin be repaid, a belief to which God's suspension of Abraham's sacrifice of Isaac might be linked (Gen 22). By displacing the death of the God with the Son's free offer of self-sacrifice and his example of virtuous life, Socinians and Milton turned the focus to the exemplary, to the kind of acts that we might emulate, and thus shortened the distance between the Son of God and the rest of us.

Newton is not entirely consistent on the atonement. He seems in several manuscripts to support the doctrine of the Son's judicial atonement, as he does in the "Drafts on the History of the Church" in speaking of "the death of the Messiah & his resurrection from the dead whereby he washed away our sins in his blood."[119] But he also stresses Christ's role as exemplar. In the "Irenicum," he writes,

> Christ by his obedience to God & particularly by his submitting to God's will even to dye an ignominious & painful death upon the Cross as an example to teach us absolute obedience in all things to God's will, has so far pleased God as to merit of him a kingdom & that God should forgive the sins of all those whom he shall chuse to be his subjects & therefore he is said to have made an attonement for us & to have satisfied Gods wrath & merited our pardon & to have washed away our sins.[120]

[118] Timothy Windsor, following William B. Hunter, has argued for the centrality of Christ's passion in *Paradise Lost*, with the three days of the War in Heaven corresponding to the three days of the Passion and Resurrection, but even in this reading, the passion is present mainly by displacement and allusion. See Timothy Windsor's "The War in Heaven: Milton's Apotheosis of Epic Catabasis," *The Journal of Religious History* 3, no. 1 (2017): 100–21; his essay draws on William B. Hunter, "The War in Heaven: The Exaltation of the Son," in *Bright Essence*, ed. Hunter, Adamson, and Patrides, 115–30.

[119] Yahuda Ms. 15.2, fol. 24^r; https://www.newtonproject.ox.ac.uk/view/texts/normalized/THEM00219, accessed March 29, 2024. See also "Theological Notebook," Keynes Ms. 2, fol 17^v.

[120] Newton, "Irenicum," fol. 36^r.

The emphasis in this passage falls more on the side of Christ as example than on Christ as judicial sacrifice. One way to read this curious passage is that the Father is moved after the fact of the passion to accept it as atonement. Moreover, the statement on the atonement is curiously oblique: "therefore *he is said to have made* an attonement" (my emphasis). In stating directly Christ's role as example and couching the atonement as if in a manner of speaking, Newton flirts with without asserting a Socinian understanding of Christ's life and death.[121]

He does, moreover, in the next breath emphasize in Socinian fashion that a free God is not bound by any obligation to substitute the Son's death for that of his creatures:

> For a man to forgive injuries without satisfaction made to him is no injustice. It's an act of mercy & more commendable then to forgive injuries upon satisfaction made. Its our duty to do so & God has in effect commanded us to do it if we expect to have our sins forgiven. And that which is an act of mercy, a duty commanded a commendable meritorious act in us cannot be injustice in God.[122]

To suggest otherwise, Newton implies, is a slur on God's goodness. And the restriction he opposes here would be a restriction on what for Newton is boundless, God's omnipotence and freedom of action. Newton's voluntarist God is not bound by Milton's rationalist God's self-imposed imperative: "Die he or justice must" (*Paradise* Lost 3.210). In their partial but radically different responses to Socinian understandings of the atonement, Milton and Newton are moved by different commitments—Milton's on the freedom that characterizes all significant creaturely action, even of the Son of God, and Newton's on the unrestricted omnipotence of the Father—that both find precedents in Socinian thought.

If significant differences with Socinian teachings make it clear that neither Milton nor Newton can be called a Socinian, it is just as clear that each held several positions in common with Socinians; they are, as I suggested above, best understood as Socinianizing in their religion. What might have drawn these two Arian thinkers to Socinian ideas? Each had a fierce commitment working out one's faith through individual study and interpretation of Scripture, a related commitment to the study of philology and the transmission history of the Bible, a merited confidence in his intellectual powers, and a disinclination to accept doctrines in conflict with reason however enshrined in opinion. They would have found all these characteristics in Socinian writers. A distinctive and salient feature

[121] Newton's vacillation on the atonement may reflect a tension between his allegiance to primitive Christianity and his interest, described in the Introduction above, in recovering the ancient ur-religion of the prisca theologia.

[122] Newton, "Irenicum," fol. 36^r.

of Socinian thought is confidence in the explicability of religious truths to reason. Habits of thought that led them to an Arian conclusion that Scripture did not support the belief that the Son was co-essential nor co-eternal with the Father would have made them susceptible to the even more heterodox Christology of the Socinians. Newton, given his strong emphasis on the absolute otherness of the omnipotent God, might have found the Socinians' emphasis on the uniqueness of the Father attractive. Milton, as I have suggested, might have found the Socinian emphasis on the Son's human nature and on the atonement as exemplary resonating with his own predilection for minimizing the distance between ontological levels, not only between humans and angels, but between humans and the divine. With their overlapping thinking about the life of matter and with their impatience with and distrust of processing Scripture through Greek metaphysics, Milton and Newton would have found allies in Socinian rejection of an immaterial and essentially immortal soul. Milton in particular resonated with Socinian emphasis on defending divine justice, and the Socinian focus on free will as the cornerstone of that defense would have been particularly congenial to one whose major poems offer narratives of the proper use of the will. Finally, both would have welcomed Socinian calls for religious toleration.

Aversion to mystery and the insistence that belief not violate reason or common sense, characteristics shared with Socinians, likely contributed to Milton's and Newton's being considered as deist or proto-deist thinkers, however much that would have appalled them. Whatever their (for them impossible) postmortem protests, their skepticism about separable immortal souls, their hylozoist understanding of matter (or extended stuff) as potentially alive, and their insistence on the adequacy of reason were all congenial to Enlightenment and post-Enlightenment ways of thinking.

Ultimately, Newton as a theological thinker, despite the daunting volume of his manuscript writings on the subject, is something of a minimalist. The wide range of significant doctrinal questions that he counts among the *adiaphora*, the things indifferent, is borne out by the flexibility and permutations of his positions. The Son is begotten before creation, but perhaps he is begotten first upon Mary. Christ's death was needed to atone for our sins, but perhaps it wasn't. Rejecting the controversial doctrinal certainties and assertions of the Nicene and Athanasian Creeds, Newton believed in a few saving truths, summed up tersely in the Apostles Creed (though even here he sometimes wavered, suggesting that there was a primitive solar religion, emphasizing one God symbolized by the fire of the sun and worshipped in vestal religions, featuring temples built around a central fire, across civilizations). What he may have found attractive in Socinianism was the emphasis on the one God the Father, omnipotent and defined by dominion, unfettered by legal requirements, but accessible to reason. His minimalism may well have been a function of his healthy respect for the vast distance between this omnipotent God and all his creatures, even his Son.

Milton, on the other hand, is more of a theological maximalist. Similarly committed to rational belief, though open to mystery when Scripture (rarely) makes it unmistakable, he is more confident that he can be decisive on matters Newton may have been content to leave undecided. In inverse relation to Newton, this may have been a function of his tendency to flatten hierarchies, to bring the divine Son closer to our experience and to our level, to make him a figure open to our successful imitation. Milton's theological ambiguities arise at the junction of doctrine and narrative, as God and the Son of God are put into action, where motivations, choices, and their consequences must play out in ways that discursive theology can leave unexamined, as I will argue in the Conclusion. Newton, as I have been arguing in this book, faced, unlike Milton, the need to make his natural philosophy square with his theology and with observed and experimental physical phenomena. Milton faced another and perhaps no less difficult challenge: the need to make his natural philosophy and theology work within the constraints of narrative literature. Each rose to the task.

Milton and Newton: Matter, God, and Modernity. Stephen M. Fallon, Oxford University Press.
 DOI: 10.1093/9780198988939.003.0005

Conclusion

Two Visionaries

William Wordsworth, a visionary himself, recalls in *The Prelude* contemplating the statue of Newton at Trinity College, Cambridge:

> I could behold
> The antechapel where the statue stood
> Of Newton with his prism and silent face,
> The marble index of a mind for ever
> Voyaging through strange seas of Thought, alone.[1]

Wordsworth's Newton has much in common with the visionary poet Milton, who describes himself as having mounted "Into the Heav'n of Heav'ns.../...and drawn empyreal air" (*Paradise Lost* 7.13–14). If Newton voyaged "alone," Milton was the poet to "see and tell / Of things invisible to mortal sight" (3.54–55), that is to say, of things invisible to the rest of us. Both saw themselves, I have argued, as prophetic figures, and this self-understanding included the sense of being chosen by God to discover and voice truths. Upon Newton's death, Alexander Pope composed a couplet epitaph: "Nature and Nature's Laws lay hid in Night: / God said, 'Let Newton be!' and all was light." Pope's Newton, like Newton's Son of God, was a mediator; where the Son as agent created the world, Newton as agent created understanding of the world. While the *Principia Mathematica* would serve for centuries as the foundation of physics and cosmology, Newton also busied himself articulating and, in some cases, forging new understandings of the life of matter, of theology, and of the Bible, areas where his thought significantly converges with and diverges from his older contemporary and fellow visionary Milton.

Milton and Newton each assumed the prophetic role of interpreter of Scripture, and their conclusions were as close to each other's as they were far from those of most of their contemporaries. While their conclusions as exegetes overlapped—the Son of God is subordinate to the Father, who is the one true God; the soul is not naturally immortal; the text of the Bible is corrupt—their differences resulted from readings driven by competing imperatives to find evidence of divine

[1] William Wordsworth, *Prelude* 3.59–63, in *The Prelude: or Growth of a Poet's Mind*, ed. Ernest De Selincourt and Helen Darbishire, 2nd rev. ed. (Oxford: Oxford University Press, 1958).

omnipotence in Newton's case and of divine justice in Milton's. And when we leave the discursive theological realm that they shared and turn to Milton's narrative poetry, we find another crucial difference between the natural philosopher and the visionary poet. Milton, in deciding to present the Father and the Son as speaking, acting characters, takes on a challenge that others, most notably Dante, declined. It is one thing to argue, as both Milton and Newton did, that the Father alone is eternal and immutable and that the Son's divinity is delegated by the Father and inferior to the Father's; it is another to show them acting together in a narrative. If Newton accepted, and overcame, the challenge of fitting mathematical theory to empirical data, Milton, unlike his fellow animist materialist writer Margaret Cavendish, accepted the challenge of fitting his imagined, created world to what might seem a counterintuitive world in which the spiritual is material.

In this Conclusion, I will take up two questions: what does a comparison of Milton and Newton tell us about the transition to a modern world? and what happens when similar bodies of thought are articulated by a visionary natural philosopher, on the one hand, and a visionary poet, on the other? I say the transition to *a* modern world, as opposed to *the* modern world. In Milton and Newton one can find important elements characterizing ways of thinking in the succeeding centuries, although it would be an overstatement to say that modern thought is a series of footnotes to either or both. A claim for prescience or influence is on the face of it more plausible for Newton than for Milton, as the *Principia Mathematica* revolutionized physics and dominated the science for over two centuries, until the intervention of Einstein and the arrival of quantum physics. When we turn, though, to theories of matter, to the status of Scripture, and to the role of reason in theology, it becomes clear that Milton and Newton are similarly pioneers.

Milton and Newton are theological outliers. Both are committed to Reformation principles of the *sola scriptura* and the priesthood of all believers. But where most Protestants accepted, for example, what was by then thirteen centuries of near consensus over the orthodoxy of a Trinity of three persons in one God, Milton and Newton with characteristic independence of mind challenged Protestant trinitarian orthodoxy as incompatible with divine revelation. Those in the late seventeenth and eighteenth centuries who read Milton's poetry and Newton's theological writing carefully raised alarms about the implications of their thought. Both were claimed by deists or suspected of deism by the more orthodox. This would have horrified either, for each was committed to a personal God continuously working in creation. But the suspicions of the wary and the enthusiasm of the deists are explicable in terms of Milton's and Newton's shared emphasis on the rational explicability of Christian doctrine and their aversion to mystery.

For both Milton and Newton, one means one, and one cannot be three, so centuries of Platonist and scholastic sophistication of the otherwise perspicuous divine communication of the fundamentals of belief had to be repudiated. From an orthodox Christian perspective, their antitrinitarianism could be viewed as a symptom of secularization, but to Milton and Newton the belief was a return to primitive Christianity and a corrective to an idolatry still tying Protestantism to the Roman Catholic Church. When Milton called in 1644 for the "reforming of Reformation it self,"[2] he was concerned about church government and the laws of marriage; by the 1660s his concern for reforming the Reformation extended to returning to primitive Christian monotheism.

Despite their strong theological commitments, Milton's and Newton's rejection of the Trinity as not only extra-biblical but irrational foreshadowed Enlightenment attacks on the irrationality of religious faith. Peter Harrison has demonstrated that the one thing many people think they know about the early Church Father Tertullian, that he wrote, "*Credo quia absurdum est*/I believe, because it is absurd," is a misquotation with an Enlightenment provenance.[3] When, in his early third-century *De carne Christi*, Tertullian wrote "*certum est, quia impossibile*/it is certain, because it is impossible," he apparently meant that reports of the death and resurrection of Christ must be true because liars would have chosen a more plausible lie, and only contemporary witnessing could have given rise to the belief.[4] Harrison traces the distortion of Tertullian's statement to the early modern period, culminating in the skeptic Voltaire's translation, "*je le crois parce que cela est absurd*/I believe because it is absurd."[5] Voltaire's repurposing of Tertullian upends the Church Father's meaning; the aphoristic saying has become a way of branding religious belief as irrational and delusionary. And Voltaire's fathering the statement on Augustine, the central Church Father, rather than the in some ways dubiously orthodox Tertullian, who had turned to Montanism and subordinationism, was a way to generalize the target to all Christianity. In this climate, the rationalist objections to trinitarian belief in Milton and Newton could seem, despite the objectors' intentions, to be hostile to Christian belief. In following Protestant principles to their logical conclusion, in taking seriously that one must work out one's belief without recourse to implicit faith in traditional doctrines, Milton and Newton might be seen, as Milton was by the deist Toland, as closer to

[2] John Milton, *Areopagitica*, *CPW* 2:553.

[3] Peter Harrison, "'I Believe Because it is Absurd': The Enlightenment Invention of Tertullian's Credo," *Church History* 86, no. 2 (2017): 339–64.

[4] For a related nineteenth-century argument that human beings would not or could not have invented the story of a God who lowered himself by being incarnate as human, see Søren Kierkegaard's *Philosophical Fragments* (1844), ed. and trans. Edna H. Hong and Howard V. Hong (Princeton: Princeton University Press, 1986).

[5] Voltaire, *Le Dîner du comte de Boulainvilliers par Mr. St. Hyacinte* (Amsterdam, 1767), 18; quoted in Harrison, "'I Believe Because it is Absurd,'" 353. In this dialogue the speaker, the Abbé Couet, attributes the saying to Augustine, most likely because Voltaire found it so attributed by Locke. Harrison lays out the transformation of Tertullian's saying in the early modern period, culminating in Voltaire, and the reasons for the perhaps strategic continuing misattribution to Augustine.

deism than to orthodox Christian faith.[6] In this way, despite themselves, the two thinkers might be viewed as participating in the erosion of faith. Like Milton and Newton, Locke would be accused of Socinianism and deism.

Milton's and Newton's shared mortalism is another instance of their early adoption of views that would characterize a modern world. Milton argued explicitly and Newton implied strongly that the body and soul are inseparable and that no naturally immortal, incorporeal soul survived the death of the body. In suggesting in the 1660s that "the distinction between these ideas [extension and thinking] will not be so great but that both may fit the same created substance, that is, but that a body may think, and a thinking being extend [*hoc est corpora cogitare vel res cogitantes extendi*],"[7] Newton anticipates Locke's argument in the *Essay concerning Human Understanding* (composed in the mid-1680s and published at the end of the decade), that it is "not much more remote from our Comprehension to conceive, that GOD can, if he pleases, superadd to Matter a Faculty of Thinking, than that he should superadd to it another Substance [i.e., incorporeal spirit], with a Faculty of Thinking."[8] In leaving behind a naturally immortal, immaterial soul, Milton, Newton, and Locke anticipated the rejection of the incorporeal soul that would come to dominate modern intellectual culture.

By the same token, they participated in the nascent vitalist conception of matter that would flower in the eighteenth century and beyond. Vitalism is a broad term. As I have used it in this book, it is not that force thought to distinguish living things from the rest of the physical universe, nor is it the Neoplatonic or Virgilian world-soul, a spirit behind and vivifying all of nature. Instead it is an "immanent vitalism," a principle of life that was seen to reside in matter itself. As clear as it was to Descartes and most of his contemporaries that an immaterial substance was necessary for thought, and as clear as it was to the Cambridge Platonists that an incorporeal substance was necessary for the initiation of any action, Milton's and Newton's sense that the principle of life, thought, and action need not be immaterial would become a dominant aspect of modern thought. As we saw in the chapter on alchemy, vitalist conceptions of life as immanent in body, as opposed to the product of an incorporeal substance superadded to matter, were not a vestige of an earlier naïve animist view being eclipsed by the rise of mechanist science. Instead it was a common feature of the thought of the new experimentalists, anatomists, and physicians.

The world taking shape in the thought of Milton and Newton is one characterized by a resistance to implicit faith in traditional doctrines; by a sense that matter, rather than being an essentially inert stuff that depends on the addition of an

[6] See John Toland's *The Life of John Milton* (London, 1699).

[7] "On the Gravity and Equilibrium of Fluids," 143 (Latin original, p. 109).

[8] John Locke, *An Essay Concerning Human Understanding*, ed. Pauline Phemister (Oxford: Oxford University Press, 2008), 345 (IV.3.vi).

immaterial substance for vitality, can itself be the site and source of life; and by an embrace of what Alexander Koyré has described as the movement from "the closed world to the infinite universe."[9] That universe is neither that limned in Pascal's unsettling and unsettled *pensée*, "the eternal silence of these infinite spaces fills me with dread,"[10] nor the nightmarish vision at the end of *Moby-Dick*'s chapter on "The Whiteness of the Whale" of a world "whose allurements cover nothing but the charnel-house within."[11] It is a living world. And if the resistance to implicit faith in received doctrines would lead to, or be feared as, ominous for religion, Milton and Newton offered models of the coexistence of (however unorthodox) religious faith with new and influential ways of thinking about the world around them.

In each chapter, I have explored the ways in which the two thinkers' surprisingly similar bodies of thought have diverged as a result of the distance between Milton's overriding commitment to defending the freedom of the will from top to bottom of the chain of being and Newton's devotion to the project of demonstrating divine omnipotence. At the end, I want to turn to another and more obvious difference, that between the visionary poet and the visionary natural philosopher. In choosing the same adjective, visionary, for the two, I am already suggesting an element of convergence belied by the anecdote of Newton's agreement with Isaac Barrow's dismissing poetry as a kind of "ingenious nonsense."

Milton and Newton were each in their own ways, to borrow William Poole's term, Restoration "world makers."[12] When Poole's world makers, working under the imperative to reconcile natural philosophy with the biblical record, strove to fit geological and cosmological evidence with the creation and flood, they had to construct hypotheses that explained that evidence in ways compatible with Scripture. And as Frédérique Aït-Touati has argued recently, there is a close link between early modern cosmological hypothesizing and fiction. She has shown that, prior to the Renaissance, astronomy was concerned with "furnish[ing] mathematical models to account for, and predict, the apparent motion of the stars. Kepler was one of the first astronomers to affirm the legitimacy of astronomical descriptions of the real nature of the cosmos, which meant bringing together its physical interpretation and mathematical (particularly geometric)

[9] Alexander Koyré, *From the Closed World to the Infinite Universe* (Baltimore: Johns Hopkins University Press, 1968).

[10] Blaise Pascal, *Pensées*, trans. A. J. Krailshemer (1966; rev. ed. Harmondsworth: Penguin, 1995), 66 (#201).

[11] Herman Melville, *Moby-Dick, or The White Whale*, Chapter 42, "The Whiteness of the Whale" (1851; reprt. Boston: St. Botolph Society, 1892), 186.

[12] See William Poole, *The World Makers: Scientists of the Restoration and the Search for the Origins of the Earth* (Oxford: Peter Lang, 2010). Poole has told a compelling story of Restoration scientists wrestling to coordinate new natural philosophical theories with the biblical record. Then as now, moving from what is observable in the present to prehistory and causation requires acts of imagination.

description."[13] Breaking with earlier, merely mathematical, geometric approaches to cosmology, Johannes Kepler as a thought experiment sent an embodied human observer to the moon, illustrating in the process that the Earth is a planet, a wandering star, like others (the Moon, Mars, Venus, etc.) in the night sky. Aït-Touati argues that in Kepler's cosmological modeling "a fiction conceived as a thought-experiment inscribes the place of matter and physics in the space of the Copernican cosmos, a space no longer merely theoretical, but concrete.... [F]iction is not opposed to knowledge but takes part in the construction of different kinds of knowledge: it confirms, anticipates, or develops them."[14] Aït-Touati contrasts the *fictional* as deployed by the hypothesizing natural philosopher with the *fictive*. The manner of hypothesizing employed by Kepler and others, including Newton, "retained a strong link with fiction conceived not as falsehood or as unreal, but as a heuristic faculty of the human imagination," as opposed to "'fictive' fictions, in the sense of illusion and falsehood."[15]

While Newton for his part from time to time insisted that he was focused on mathematical description of celestial and terrestrial phenomena rather than speculating on physical or immaterial causes, and while he famously asserted in the 1713 "General Scholium" to the *Principia Mathematica* that, "*Hypotheses non fingo* (I do not frame/feign hypotheses)," he does throughout his career frame hypotheses. I have alluded in this connection to his evolving thinking about gravity. Sometime in the mid-1660s, Newton speculated on the motion of the material agent causes of gravity in his manuscript notebook headed "Questiones Quædam Philosophiæ":

> The matter causing gravity must pass through all the pores of a body. it must ascend againe. for else the bowells of the earth must have had large cavitys...to conteine it in, 2 or else the matter must swell it. 3 the matter that hath so forcibly borne {down} the earth & all other bodys to the center...cannot if added to gether be of a bulke so little as the earth, for it must descend exceding fast & swift as appeares by the falling of bodys, & exceeding weighty pressure to the Earth. It must ascend in another forme than it descendeth or else it would have a like force to beare bodys {illeg} that it hath to press them downe & so there would bee no gravity.[16]

A decade later, in his December 1675 "Hypothesis explaining the properties of light," he again speculates on the cause of gravity, as we have seen in the chapter in alchemy; at that time he imagines that gravity results from corporeal aethereal

[13] Frédérique Aït-Touati, *Fictions of the Cosmos: Science and Literature in the Seventeenth Century*, trans. Susan Emanuel (Chicago: University of Chicago Press, 2011), 37.

[14] Aït-Touati, *Fictions of the Cosmos*, 30, 36.

[15] Aït-Touati, *Fictions of the Cosmos*, 197.

[16] Newton, "Questiones quædam philosophiæ," Cambridge University Library MS Add. 3996, 97^{r}; https://www.newtonproject.ox.ac.uk/view/texts/normalized/THEM00092, accessed December 20, 2023.

spirits that, having risen, then condense and descend and exert pressure driving objects downward.[17] By the turn of the century, Newton would leave behind mechanical hypotheses on gravity, and in the Queries to the Opticks Newton hypothesized that inasmuch as gravity acted at a distance, and inasmuch as it penetrated to the interior of material bodies, it must not be grossly material, and moreover, that the cause of gravity was the will of God.[18]

In insisting "*Hypotheses non fingo*," Newton claimed to be limiting himself to mathematical description and prediction. But in his changing hypotheses for the causes of gravity, Newton was attempting a "physical interpretation" of "the real nature of the cosmos." In doing so, Newton was imagining, constructing likely stories, heuristic fictions, of an unknown force that he had measured mathematically. He is straddling two discourses, in one of which he is following an ancient model of "saving the phenomena," according to which astronomical hypotheses were merely useful fictions not purporting to make truth claims, and the new Keplerian attempt to imagine the physics, and not merely the mathematics, of the universe. Aït-Touati has argued that early modern astronomical hypothesizing was a key meeting point of what we now bifurcate as the separate discourse of "science" and "literature."[19]

If Newton had something in common with the poets he ostensibly relegated to mere fantasizing about the unreal, there remains a crucial difference from a poet like Milton who also imagined and described a physical universe. Newton had to make the mathematics and the concomitant predictions work. Whatever the cause of gravity—whether the exertion of an immaterial power by God or the rising, coalescing, and descent of vital aereal niter—his proposal that gravitational force is inversely proportional to the square of the separation distance between the objects had to fit the astronomical data; the mathematics had to work.

Milton, despite his strong amateur's interest in mathematics, and despite what Catherine Gimelli Martin, Dennis Danielson, and John Leonard have demonstrated about his adoption of the new Copernican astronomy, did not face the same challenge in imagining the physical universe.[20] He does demonstrate astronomical sophistication in describing the journeys of Uriel, stationed on the sun, to and from the Garden. Milton describes his first flight thus: "Thither came

[17] Newton, "Hypothesis explaining the properties of light," in Thomas Birch, *The History of the Royal Society*, vol. 3 (London: 1757), 3:249–51.

[18] John Henry, "Newton, the Sensorium of God, and the Cause of Gravity," *Science in Context* 33 (2020): 329–51; see p. 342; https://doi.org/10.1017/S0269889721000077; accessed January 30, 2024.

[19] Aït-Touati, *Fictions of the Cosmos*, 6.

[20] See Catherine Gimelli Martin, "'What If the Sun Be Center to the World': Milton's Epistemology, Cosmology, and Paradise of Fools Reconsidered," *Modern Philology* 99 (2001): 231–65; Dennis Danielson, "Astronomy," in *Milton in Context*, ed. Stephen B. Dobranski (Cambridge: Cambridge University Press, 2010), 213–25; and John Leonard, *Faithful Labourers: A Reception History of Paradise Lost, 1667–1970*, 2 vols (Oxford: Oxford University Press, 2013), 2: 705–19.

Uriel, gliding through the *even* / On a sunbeam" (*PL* 4.555–56; my emphasis). Thomas Newton unlocked the meaning of "even," arguing that Uriel, riding on a sunbeam, traveled the fastest way, when the sun's rays were level at evening with the gate of Eden; a few lines before, Milton had observed that at that moment "the setting sun / Slowly descended, and with right aspect / Against the eastern gate of Paradise / *Leveled* his evening rays" (4.540–43).[21] The level, non-inclined beam between sun and gate was the shortest beam in the diurnal rotation of the earth. In his return flight, Uriel is assisted by gravity, as the sun has fallen far in the western sky over the mid-Atlantic Azores, below the level of the gate:

> Uriel to his charge
> Returned on that bright beam, whose point now raised
> Bore him slope downward to the sun now fall'n
> Beneath th' Azores; whether the prime orb,
> Incredible how swift, had thither rolled
> Diurnal, or this less voluble Earth
> By shorter flight to th' east, had left him there
> Arraying with reflected purple and gold
> The clouds that on his western throne attend. (4.589–97)

Here, as is the case in Book 8 when Raphael answers Adam's cosmological query, Milton begins with the Ptolemeic, geocentric model before turning to a Copernican heliocentric one. The order is significant, for reasons that Raphael will make clear. The first in each case is earthbound, governed by what is apparently the case for the human observer. Raphael begins with a remark on the geocentric model, "*though so it seem* / To thee who hast thy dwelling here on Earth" (8.117–18; my emphasis), before shifting to the Copernican model, which has the virtue of addressing Adam's skeptical question about the speed of the "restless revolution day by day" required by the geocentric perspective (8.31) and offering a more plausible explanation of the succession of day, night, and stars as caused by the smaller rotation of the Earth. In describing Satan's landing in the Paradise of Fools during his flight from Hell to Earth, Milton seems for a moment to provide his universe with Ptolemy's concentric planetary spheres and the fixed sphere of the Primum Mobile, but as the Jonathan Richardsons, father and son, observed in 1734, these spheres "are no more parts of *Milton's* System of the new Creation than the Wicket Gate in the next line.... These were Some of the same Reveries of the Philosophers and Astronomers, *Quaint Opinions* to be Laught at, as VIII.78. he says This by flinging them into the Paradise of Fools."[22]

[21] See Thomas Newton's notes to *Paradise Lost* 4.541 and 4.590 in *Paradise Lost, The Seventh Edition* (London, 1770), 300, 304.

[22] Quoted in Leonard, *Faithful Labourers*, 2:707.

Milton was in these ways attentive to natural philosophy in squaring the setting of his epic to emerging ideas in cosmology and astronomy. But the squaring of his animist materialism with the universe of his epic presented a greater challenge, one he did not skirt as Margaret Cavendish had in her narrative published a year earlier in 1666. In the letter "To the Reader" prefacing *The Blazing World*, she distinguishes the roles and responsibilities of the natural philosopher and the writer of imaginative fiction:

> *fictions* are an issue of man's fancy, framed in his own mind, according as he pleases, without regard, whether the thing he fancies, be really existent without his mind or not; so that reason searches the depth of nature, and enquires after the true causes of natural effects; but fancy creates of its own accord whatsoever it pleases, and delights in its own work. The end of reason, is truth; the end of fancy, is fiction.[23]

Committed to writing a true epic, Milton did not leave himself the option of separating natural philosophy and narrative. The strain shows itself in narrative details, particularly in dealing with the bodies of Satan and the other fallen angels. If, as Raphael explains, bodies are "more refined, more spiritous, and pure, / As nearer to him [God] placed or nearer tending" (5.475–76), why is Satan, who is able, for example, to shapeshift and enter the serpent, more tenuously corporeal than the unfallen Adam and Eve? If this is a discrepancy, it can be rationalized in two ways. First, the focus of Raphael's words is on dynamic movement: one becomes more spiritous than one was by moving closer to God, and by the same token less spiritous or more dense as one moves away from God. There is ample evidence of Satan's hardening relative to his former state after his rebellion. Second, Raphael specifies that this ontological motion unfolds "in bounds / Proportioned to each kind." Reaching the lowest limits of angel embodiment may still leave Satan more tenuously embodied that Adam and Eve currently are, though presumably not more so than Adam and Eve might eventually have become, had they remained obedient and experienced their "bodies…at last turn[ing] all to spirit, / Improved by tract of time, and winged ascend[ing] / Ethereal" as the unfallen angels' (5.497–99).

Other moments in the narrative cannot be as easily rationalized, as Milton treats Satan's body in different ways for different rhetorical and thematic

[23] Margaret Cavendish, *The Blazing World and Other Writings*, ed. Kate Lilley [1992; rpt. Harmondsworth: Penguin, 2004], 123. I contrast Milton's attempted fidelity to his metaphysical model in his narration of *Paradise Lost* with Cavendish's intentional and strategic violation of her own in the narration of *The Blazing World*, in "Imagining Animist Materialism in *Paradise Lost* and *The Blazing World*," in *Margaret Cavendish and John Milton*, ed. Lara A. Dodds, Ann Coiro, and Lisa Walters, forthcoming from Oxford University Press. For a related argument, see Stephen Hequembourg, "The Poetics of Materialism in Cavendish and Milton," *SEL: Studies in English Literature 1500–1900* 54, no. 1 (2014): 173–92.

purposes. As noted a moment ago, Satan is sufficiently tenuous to enter and share the space of the serpent without waking the "subtlest beast": "in at his mouth / The Devil entered.../...but his sleep / Disturbed not" (9.187–91). On the other hand, unable or disinclined to pass through the tangled undergrowth guarding entrance to the Garden, he instead leaps over the apparently impenetrable vegetation, as Milton offers the morally charged double metaphor of Satan as wolf and thief:

> when th' arch-felon saw
> Due entrance he disdained, and in contempt,
> At one slight bound high over leaped all bound
> Of hill or highest wall, and sheer within
> Lights on his feet. As when a prowling wolf,
> Whom hunger drives to seek new haunt for prey,
> Watching where shepherds pen their flocks at eve
> In hurdled cotes amid the field secure,
> Leaps o'er the fence with ease into the fold:
> Or as a thief bent to unhoard the cash
> Of some rich burgher, whose substantial doors,
> Cross-barred and bolted fast, fear no assault,
> In at the window climbs, or o'er the tiles;
> So clomb this first grand thief into God's fold. (9.179–92)

It serves Milton's rhetorical purpose to have Satan overleap a wall like a thief into a house or a wolf into a sheepcote, though it is not clear why he is unable or unwilling pass through the obstacle as he passes through and into the body of a snake. On Newton's surmise that gravity's force might need to be immaterial to penetrate to the center of gravitating bodies, Satan would have to be immaterial or all but immaterial. On the other hand, by the principle of bodily density as a measure of distance from God, he must be significantly densely bodied, as, after all, he was subject, unlike the loyal angels, to bodily injury in the War in Heaven, when the sword of Michael "deep ent'ring shared [sheared] / All his right side" (6.326–27). By the same token, the newly physically dense devils are trapped for a time in their crushed armor:

> Their armor helped their harm, crushed in and bruised
> Into their substance pent, which wrought them pain
> Implacable, and many a dolorous groan,
> Long struggling underneath, ere they could wind
> Out of such prison, though spirits of purest light,
> Purest at first, now gross by sinning grown. (6.656–61)

Note the marvelous meeting of pun and aural repetition in these lines, as the "dolorous groan" is repeated in the devils' new state, "now gross by sinning grown/[groan]."

Milton, as we have seen, constructs the world of *Paradise Lost* in accordance with the principles that all that is, is material and alive, and that the soul is not a separable, immaterial substance. But under the demands of narrative and the elaboration of figurative meanings he compromises the natural philosophy in *ad hoc* ways. While Newton may offer differing hypotheses over time, he is always at any moment constrained to be consistent in his understanding and articulation of his natural philosophy. One might conclude from this key difference that Milton's task is significantly easier than Newton's, as, like Philip Sidney's poet, Milton is "not enclosed within the narrow warrant of her [Nature's] gifts, but freely ranging only within the zodiac of his own wit."[24] He is a poetic visionary, imagining how the world both is and might be, not what John Donne called "Nature's Secretary, the Philosopher."[25]

But if the natural philosopher faced the challenges of squaring arguments with the observable, measurable phenomena and of strict internal consistency, narrative poets faced challenges of their own. Discursive formulations are tested and stressed when put into motion and narrative. And if Newton faces a more difficult task in grounding his vitalist natural philosophical positions in observed phenomena than does Milton in describing the world in his poems, Milton faces the more difficult task in the area of shared theology: How does one narrate the relations between an eternal, immutable Father and his non-eternal, non-coessential Son? And how does one narrate the career of an incarnate Son of God, a being with delegated divinity born into the world as a man? Ultimately, the challenge for the poet may be greater than that for the natural philosopher, for an adequate articulation of a natural philosophy embracing and consistent with empirical observation and mathematics is at least theoretically possible, but the visionary poet's attempt to write the "word and effectual might" of an immutable God the Father into a linear narrative must ultimately run into the dilemma of presenting that God as acting in time, and thus participating in change.[26]

Milton and Newton agree that the eternal, immutable Father is the one God, who alone possesses in himself omnipotence, omniscience, and other divine attributes, and that the Son was begotten in time, that he is an instrument of the Father, and he possesses divinity and divine attributes not in himself but by delegation from the Father. Milton is highly unusual among early modern authors of

[24] Sir Philip Sidney, *A Defence of Poetry*, in *The Miscellaneous Prose of Sir Philip Sidney*, ed. Katherine Duncan-Jones and Jan van Dorsten (Oxford: Oxford University Press, 1973), 78.

[25] John Donne, Satyre I, l.6, in *John Donne: The Satires, Epigrams and Verse Letters*, ed. W. Milgate (Oxford: Oxford University Press, 1967), 3.

[26] In the following pages I am indebted to Samuel Fallon's "Milton's Strange God: Theology and Narrative Form in *Paradise Lost*," *ELH* 79 (2012): 33–57.

divine heroic poems in presenting the Father as a speaking character, and thus involving him as an actor in a narrative. It is in large part for this reason that Andrew Marvell's initial concern upon picking up *Paradise Lost* was that Milton "would ruin.../ The sacred truths to fable and old song."[27] As Samuel Fallon has observed, "One can make certain claims about God, but transforming those claims into plausible accounts of how it all actually works turns out to be more difficult."[28] As Fallon demonstrates, in their dialogue in Book 3, the Son draws the Father into engagement with the angels and into narrative, by countering the Father's initial discursive, hermetically logical account of the impending fall and possible redemption of humankind with a story casting the Father as an intending actor in time. Fallon shows how the unfolding narrative, with the immutable Father's call for a volunteer to do what an immutable God by essence and definition *cannot* do—die for humankind—illustrates the paradoxical impotence of the Father, who must depend on another to provide the solution reconciling justice and mercy. As Fallon puts it,

> The more Milton shows God speaking and acting, the more apparent become the things he cannot do: he cannot die; he cannot suffer; ultimately, he cannot change. Whereas Adam, Eve, Satan, and even the Son are all capable of transformation—as demonstrated by the two Falls on the one hand and the Son's ascent "[b]y merit more than birthright" on the other—the Father is bound by his nature to always remain the same (*PL*, 3.309). In theological terms, immutability is an aspect of perfection, rendering God invulnerable to contingency. But for a poet writing God, it has the curious side effect of making him, in a fundamental sense, unequal to the demands of narrative.[29]

There is a reason that other poets chose not to present the Father as a speaking character. As one attempts to narrate an immutable, eternal figure acting in time, one reaches an impasse. As Fallon concludes succinctly, in Milton's epic "God fails narrative as much as narrative fails God."[30] We can create and understand discursive articulations of God, but at the same time there is a desire to understand how that God interacts with us and the rest of creation. For this, we turn to story, to imagining God as doing something the discursive formulation forbids, that is, acting in time. If the narration ultimately fails to bridge the gap between itself and the theological, discursive definition of God, the failure can nevertheless be a noble attempt to come as close as possible to closing that gap, to understanding the finally incomprehensible. In this, it shares something

[27] Marvell, "On Mr. Milton's *Paradise Lost*," 7–8, MLM 287.
[28] Fallon, "Milton's Strange God," 33.
[29] Fallon, "Milton's Strange God," 45–46.
[30] Fallon, "Milton's Strange God," 52.

with the attempt of theodicy itself. If Milton, given the imperative of realist narrative that actions be explicably motivated rather than random, and given further his readers' knowledge from Genesis that Adam and Eve will fall, may not have finally succeeded in demonstrating his "great argument" that Adam's and Eve's falls were freely willed, I propose that he came as close as one could come.[31]

Where Fallon focuses on the relation of the Father and the Son, I want to turn attention to the relation between the Son as instrument and substantial expression of the Father and the Son as a separate actor, or in other words the relation of the Son as passive instrument and the Son as agent. The epic is studded with formulations of the Son as visible manifestation of the ineffable Father, as, for example, the "radiant image of [the Father's] glory" (3.63). In several of the passages, Milton repeats the phrase "all his Father": "in him *all his Father* shone / Substantially expressed" (3.139–40); "he *all his Father* full expressed" (6.720); "*all his Father* in him shone" (7.196); "he full / Resplendent *all his Father* manifest / Expressed" (3.139–40, 6.720, 7.196, 10.65–67; my emphasis). I will suggest that the relation is not entirely reciprocal. The Son may visibly express "all his Father," but the Father, paradoxically, may not direct all that the Son does.

The difference arises in the tension between the Son's acting as instrument of the Father and the Son's acting on his own accord as agent. The Son, the Father declares, is "alone / My Word, my wisdom, and effectual might" (3.169–70). The angels in their hymn at the end of the Dialogue in Heaven in Book 3 praise the Son as both the visible expression of the Father and the instrument of the Father's will at the Creation and in the War in Heaven:

Thee next they sang of all creation first,
Begotten Son, divine similitude,
In whose conspicuous count'nance, without cloud
Made visible, th' Almighty Father shines,
Whom else no creature can behold; on thee
Impressed the effulgence of his glory abides,
Transfused on thee his ample spirit rests.
He Heav'n of Heav'ns and all the Powers therein
By thee created, and by thee threw down
Th' aspiring Dominations: thou that day
Thy Father's dreadful thunder didst not spare. (3.383–93)

[31] See my "Narrative and Theodicy in *Paradise Lost*," *Milton Studies* 61 (2019): 40–64.

As in the *De Doctrina Christiana*, the Father creates by means of the Son, and here as well it is, at least from the angels' perspective, the Father who, using the Son as instrument, "threw down" the rebel angels. The narrative endorses the angels' sense that the Son is the Father's instrument in the middle of the epic. Having determined to create the world, the Father sends the Son as instrument:

> Thou my Word, begotten Son, *by thee*
> This *I* perform, speak thou, and be it done:
> My overshadowing Spirit and might with thee
> I send along, ride forth, and bid the deep
> Within appointed bounds be heav'n and earth. (7.163–67; my emphasis)

A book earlier, the Father likewise sends the Son out, as an instrument of his will, to defeat the rebel angels, beginning his charge thus:

> Effulgence of my glory, Son Beloved,
> Son in whose face invisible is beheld
> Visibly, what by deity I am,
> And in whose hand what by decree *I* do. (6.680–83; my emphasis)

The syntax is complex but the sense is clear: in the Son's *hand* is visible what "by decree I [the Father] do." The Son here eagerly accepts his role as instrument of the Father's will:

> this I my glory account,
> My exaltation, and my whole delight,
> That thou in me well pleased, declars't thy will
> Fulfilled, which to fulfill is all my bliss. (6.726–29)

In these lines Milton's Son appears to be entirely submerged in his identity as the visible manifestation and instrument of the ineffable and immutable Father, enacting in time the eternal Father's will.

This picture of a passive Son putting into effect the will of a superior Father fits the Arian understanding of God, though in practice, as will become clear in a moment, it is closer to Newton's than Milton's Arianism. Milton and Newton, as we have seen, argued that the Father and Son are one not in essence but in union of wills. A salient moment of the subsumption of the Son in the Father occurs when the Father sees "Rebellion rising" after his exaltation of the Son in the chronologically earliest moment in the epic (5.715). The Father's words to the Son suggest real anxiety about the challenge:

> Nearly it now concerns us to be sure
> Of our omnipotence, and with what arms
> We mean to hold what anciently we claim
> Of deity or empire....
> ...
> Let us advise, and to this hazard draw
> With speed what force is left, and all employ
> In our defense, lest unawares we lose
> This our high place, our sanctuary, our hill. (5.721–32)

The Son's response, "Mighty Father, thou thy foes / Justly hast in derision, and secure / Laught'st at their vain designs" (5.735–37), with its acknowledgment of a dismissive sarcasm far from obvious from the Father's words, suggests an intimacy and even a unity of mind with the Father.

While in the discursive prose of the *De Doctrina Christiana* Milton's position on the instrumentality of the Son is indistinguishable from Newton's, as he thinks through the matter in narrative poetry, a gap begins to open. This gap opens particularly as Milton narrates the interaction of Father and Son about the possibility of the redemption of fallen humankind. In the theological treatise, Milton offers this formulation of redemption: "**Redemption** is the act by which, in accordance with the eternal counsel and grace of God the Father, Christ, sent in the fullness of the times, redeemed all believers at the cost of his own blood, paid in full [and] voluntarily" (*DDC* I.14; *OCW* 8.1.469). While indicating that the Son pays the cost "voluntarily," he also specifies here that he does so "in accordance with the eternal counsel...of God the Father." The Father decrees redemption from eternity, and at a certain point in time the Son is sent to fulfil that decree. When processed through narrative, however, the picture changes significantly, in a way that suggests (1) that the Son possesses his own independent will and real agency, as opposed to serving at all times as a non-autonomous instrument of the Father, and (2) that Milton is *thinking* through his poetry as opposed to merely versifying the theology of the treatise. And as he thinks through his poetry, Milton wrestles with the challenge of narrating the words and actions of a Son who is, simultaneously, an expression of the Father and an independent agent. Adding to the complication of the Son is the fact that, given that he is the Father's Word and that the ineffable Father is inaudible, the words of the Father must ultimately, despite the narrative fiction of a dialogue of two audible actors, be spoken by the Son, who then replies with his own words.

As the dialogue of Father and Son in Book 3 commences, the Father predicts the fall of humankind, insists that the fall results from creatures' free will and not a divine predestinating decree, and that he will nevertheless share grace with "man deceived / By the other [Satan] first" (3.130–31). In replying, the Son does more than reiterate and endorse the will of the Father; he makes a new argument for what the Father has tersely determined to do.[32] To condemn humankind would raise questions about the Father's justice: "That be from thee far, / That far be from thee, Father, who art judge / Of all things made, and judgest only right" (3.153–55). It would, the Son continues, give Satan the victory. If the Father were to destroy Creation, a move not suggested in the Father's first speech, "So should thy goodness and thy greatness both / Be questioned and blasphemed without defense" (3.165–66). This is a curious moment, as the Son seems to labor to persuade the Father toward a course the Father has already chosen. The Father says as much, when he replies, "All hast thou spoken as my thoughts are, all / As my eternal purpose hath decreed" (3.171–72). In a moment of unnecessary persuasion, of preaching to a choir of one, a sliver of daylight opens between the Father and the Son. The Son adds that the Father's failure to bestow grace and mercy would leave the Father open to criticism and even blasphemy.

That sliver widens in the next exchange. The Father expands on his merciful plan to offer fallen humans grace but insists that Justice requires a "rigid satisfaction, death for death," and concludes by addressing the assembled angels and the Son, "Which of ye will be mortal to redeem / Man's mortal crime, and just th'unjust to save, / Dwells in all Heaven charity so dear?" (3.212, 214–16). This is an extraordinary moment, one without a place in the *De Doctrina Christiana*'s chapter on the redemptive office of the Son (I.14), where the dispatching of the Son to die for humankind is part of the "eternal counsel." As Milton thinks through in narrative form the Son's willingness to suffer and die for humankind, he imagines a scene in which the Father does not direct the Son to assume the redemptive role but asks for a volunteer from the assembled thousands. A period of silence follows the call (3.217–24), suggesting a space for intense introspection and soul searching in the assembly, before the Son offers himself. One might suggest that the Father and the Son both knew, prior to the moment (the Father eternally and the Son since his begetting), that one would call and the other would respond heroically. The problem is that this would elide the difference between the Council in Heaven and the Council in Hell in the preceding book. In both there would be a regal figure calling for a volunteer and in both there would be a pregnant, engineered pause before the volunteer steps forward. If the Father and Son are following a script that both knew beforehand, the Council in Heaven is as tawdry, spurious, manipulated and manipulating as the Council in Hell, a charade to impress a gullible audience.

The terms of the Son's volunteering and the Father's response make clear that this is a real as opposed to a scripted and stage managed exchange. After offering

[32] Samuel Fallon, "Milton's Strange God," 43–44.

himself ("Behold me then, me for him, life for life / I offer, on me let thine anger fall" [3.236–37]), the Son contemplates what his death will entail:

> Though now to Death I yield, and am his due
> All that of me can die, yet that debt paid,
> Thou wilt not leave me in the loathsome grave
> His prey, nor suffer my unspotted soul
> Forever with corruption there to dwell;
> But I shall rise victorious. (3.245–50)

The phrase "Thou wilt not leave me" hovers between the indicative and the imperative, as the Son at once seems sure and seeks (or even demands) assurance that the Father will not "leave [him] in the loathsome grave," Death's "prey." The latter reading, which suggests some real distance between the Son's perspective and the Father's, some potential daylight between their wills, seems all the more plausible as Milton narrates the pause between the Son's speech and the Father's reply. The Son, the narrator observes, "attends the will / Of his great Father" (3.270–71). The Son awaits the Father's response; he finishes his speech expecting the Father to endorse or reject what he has said.

The Father, significantly, responds in a markedly different manner than he had to the Son's first speech in the dialogue, when he had said "All hast thou spoken as my thoughts are." Instead, the Father now seems animated, grateful, and relieved: "O thou in Heav'n and Earth the only peace / Found out for mankind under wrath, O thou / My sole complacence!" (3.274–76). (The exclamation point appears in both the first and second editions.) In Milton's imagination of the scene, this is a turning point for the Father in his relation to and in his understanding of the Son:

> Because thou hast, though throned in highest bliss
> Equal to God, and equally enjoying
> God-like fruition, quitted all to save
> A world from utter loss, and hast been found
> By merit more than birthright Son of God,
> Found worthiest to be so by being good,
> Far more than great or high; because in thee
> Love hath abounded more than glory abounds,
> *Therefore* thy humiliation shall exalt shall exalt
> With thee thy manhood also to this throne;
> . . .
>   all power
> I give thee, reign forever, and assume
> Thy merits; under thee as Head Supreme

> Thrones, Princedoms, Powers, Dominions I reduce:
> All knees to thee shall bow, of them that bide
> In Heaven, or Earth, or under Earth in Hell. (3.305–22; my emphasis)

The Son's birthright is his status as the only begotten Son of the Father. It is in that capacity that he is the visible expression of the Father and that he shares the Father's will and serves as the instrument of that will. It is the free offer of himself, according to his own will, that leads to his new distinction, to be by *merit* the Son of God. Having already, in the exaltation that triggered a rebellion that led ultimately to the temptation and fall of humankind, appointed the Son as "head" of the angels who should henceforth "confess him Lord" (5.606, 608), the Father only now gives the Son "all power," promises that he will "reign forever," and directs the angels to "Adore the Son, and honor him *as me*" (3.343; my emphasis).[33]

In a way that does not appear (nor does it need to appear) in the discursive formulations of the *De Doctrina Christiana* (or in Newton's theological writings), the Son by his agency and not merely as instrument of the Father affects, and arguably directs, the course of salvation history. It may be significant that it is in the chapter on the office of redemption that Milton, citing Christ's address to the Father in the Garden of Gethsemane in Matthew 26:39 ("not as I will, but as you") speaks of the Son's will being separate from the Father's; the verse "does not imply a twofold will unless he [the Son] is the same as the Father; which…cannot be admitted" (*DDC* I.14; *OCW* 8.1:485). Taking on the task of narrating sacred history required Milton, in a way not required of his fellow Arian Newton, to strenuously work out and elaborate the relation of the Father and the Son and, moreover, the relation of Son as independent agent and the Son as manifestation and instrument of the Father.

Milton faces an even larger challenge in *Paradise Regained*, where the Son has a dual nature. For Newton, as we have seen, the incarnate Christ had a single, divine nature. For the Socinians, whose thinking Milton in some ways approximated, the incarnate Christ had a single, human nature; but for the poet, divine and human natures are joined in the person of Jesus. As we have seen, Milton does not claim to know *how* the person of Jesus can contain two natures, but he is convinced by Scripture *that* he does so. In the brief epic, with its extensive

[33] What is missing in the extraordinary exchange of Father and Son in Book 3 are expressions of love for each other. There is the Son's love for humankind ("his meek aspect / Silent yet spake, and breathed immortal love / To mortal man" (3.266–68), and the Father lauds the Son's offer to sacrifice himself as evidence that "Love hath abounded more than glory abounds" (3.312), but the Son has acted to preserve the Father's honor, and the Father responds by granting power to the Son. The Father does name the Son "My sole complacence" (3.276), but the *OED* offers as the first meaning "complacence" a "Pleasure or satisfaction in one's own condition or doings," and offers this phrase from *Paradise Lost* as the original source of the meaning "An Object or source of pleasure and satisfaction" (2.c). The bountiful giving and receiving of love in the persons of the trinity in, for example, St. Bonaventure, is not a feature of the dialogue.

imagining of not only the spoken words and actions but also the internal musings of the Son of God, Milton must find a way to make a plausible narration centered around a person with a mysteriously dual nature. The difficulty of the task is articulated in the poem's invocation:

> inspire,
> As thou art wont, my prompted song else mute,
> And bear through highth or depth of nature's bounds
> With prosperous wing full summed to tell of deeds
> Above heroic, though in secret done,
> And unrecorded left through many an age,
> Worthy t' have not remained so long unsung. (*Paradise Regained* 1.11–17)

Given the plot's sources in the Gospels, what does Milton mean by "unrecorded"? In contrast to Marco Giralamo Vida's 1535 six-book epic spanning the life of Christ, the *Christiad*, Milton concentrates, and vastly expands, one episode in that life, the temptation in the desert. As was the case with *Paradise Lost*, Milton adds incident, dialogue, and internal musings not present in the sources, sixteen verses in Matthew (Matt 3:13–4:11) and fifteen verses in Luke (Luke 3:21, 4:1–14). The events to be recorded were "in secret done," in two senses: not only did they not find their way into the Gospels and thus into the Christian story, but also most of the action unfolds internally and in unobserved dialogue. Here, without the armature of Gospel story and orthodox narrative, Milton, the visionary poet, explores and articulates the consciousness of the God-Man Jesus.

A recurring motif in the opening pages of *Paradise Regained* is the humanity of the Son of God. The narrator describes the Son—whom an envious Satan has heard pronounced "beloved Son" by the "Father's voice"—as "th' exalted *man*" (1.36; my emphasis). (That voice, of course, as ineffable, must be mediated by the Holy Spirit, by an angel, or by the Son himself.) The Father informs Gabriel that he "can produce *a man* / Of female seed, far abler to resist" Satan's temptations than even Job (1.150–51). Milton navigates a course between the Arian, according to which the pre-existent quasi-divine Son is incarnate in human form, and the Socinian, in which a human being is raised to divine status for his virtues. On the one hand, Milton endorses the idea, both Arian and orthodox, that the Son pre-existed his incarnation as Jesus and served as the Father's instrument in creation, but on the other he hints at the incarnate Son achieving quasi-divine status by his actions and virtues. In the same speech to Gabriel, the Father alludes to a test that the Son must pass: "this man born and now upgrown, / To show him worthy of his birth divine / And high prediction, henceforth I expose / To Satan" (1.140–43). As with the call for a volunteer in Book 3 of *Paradise Lost*, if this is merely stage-managed, a test with a foregone conclusion, the stakes of the test and the heroism of the Son evaporate. The Father insists that the test has been prepared

> That all the angels and ethereal powers,
> They now, and men hereafter may discern,
> From what consummate virtue I have chose
> This *perfect man, by merit called my Son.* (1.163–66; my emphasis)

If the conclusion of the test were foregone, because the quasi-divine, only begotten Son of the Father was impervious to temptation, it would be a matter of birthright, of divine DNA unshared by the human Jesus. This is a second moment, following from the Son's offer to die in Book 3 of *Paradise Lost*, when the Son will be "found / *By merit more than birthright* Son of God" (*Paradise Lost* 3.308–9). The Son can be a "perfect man" because of his human nature, as Milton's acceptance of the hypostatic union would require, a nature as authentic as his (subordinate) divine nature.

Milton finesses the two natures of the Son in his narration of the Son's early years and his growing consciousness of his special status. The Son recalls his own experience of an intellectually and spiritually precocious childhood: "When I was yet a child, no childish play / To me was pleasing, all my mind was set / Serious to learn and know" (1.201–3). Where the Gospel of Luke presents Jesus "listening to [the rabbis] and asking them questions" (2: 46), Milton's Jesus "propose[s] / What might improve my knowledge or their own" (1.212–13). This is not merely a Son with a human mind, but one with a mind remarkably like, and arguably patterned on, Milton's recollection of his own youth. It is by reading and interpreting Scripture, and by the testimony of Mary, not by direct access and conscious access to his divine nature, that the Son begins to understand that he is not the merely human son of Joseph. Mary's testimony sends him back to Scripture:

> This having heard, straight I again revolved
> The law and prophets, searching what was writ
> Concerning the Messiah, to our scribes
> Known partly, and soon found of whom they spake
> I am. (1.259–63)

In the Son's "I am" Milton alludes to the Tetragrammaton, the name of God, "I am that I am." This episode is carefully layered. Exploratory reading of the Hebrew Scriptures against the background of lived experience is supplemented in turn by the direct testimony from Mary, recounting the Greek Testament's story of the Annunciation, and the return for confirmation to the Hebrew Scriptures before conviction.

But the new understanding is not complete; the Son still has much to learn about what it means to be the son of the "Eternal King" mentioned by Mary, and what it means to be the "Messiah" of the Hebrew Scriptures (1.236, 261). The human consciousness and will are not eclipsed by the divine consciousness and will. The Son at the outset acknowledges the limits of his knowledge and understanding. He is led by the Spirit into the wilderness "to what intent / I learn not yet, perhaps I need not know; / For what concerns my knowledge God reveals" (1.291–93). The ensuing dialogue with the tempter Satan is an exercise in biblical interpretation and, on Satan's part, manipulation. And, like the Father, the Son views his life as one of real, not sham or scripted, trial:

> What if he [the Father] hath decreed that I shall first
> Be tried in humble state, and things adverse,
> By tribulations, injuries, insults,
> . . .
> just trial ere I merit
> My exaltation without change or end. (3.188–97)

Offered glory by Satan, the Son replies, "This is true glory and renown, when God / Looking on the earth, with approbation marks / The just *man*" (3.60–62; my emphasis).

Through the middle books of the brief epic, the Son hovers between an increasingly enlightened human consciousness and a dawning divine consciousness. The two come into tantalizingly close contact when Satan taunts the Son, who is not anxious to attain glory, with the biblical record of the Father's desire for glory. The Son replies, "And reason, since *his word* all things produced" (3.122–23; my emphasis). The Son at this stage does not yet know that he is himself this "word [that] all things produced," and he will not know this until the end of the epic, when the angels celebrating his victory over temptation allude to his preexistence "throned / In the bosom of bliss, and light of light / Conceiving" (4.596–98). While we can discern a Socinian adoptionist strain in the career of the Son in *Paradise Regained*, the brief epic is at the same time an Arian story of the Son of God in his human incarnation coming to the knowledge and memory of his Arian pre-history in Heaven, a history of reigning as vicegerent, of defeating Satan in the War in Heaven, and of creating the universe.[34]

Before this moment, the Son is wrestling with the conundrum that Satan raises. While both know that the Son in the brief epic has been named beloved Son by the voice of God, Satan observes that "sons of God both angels are and men" and wonders if the Son "in higher sort /...bears't that title" (4.197–99). There is at least one moment well into the brief epic when the Son seems far from

[34] John Rogers, "*Paradise Regained* and the Memory of *Paradise Lost*," in *The Oxford Handbook of Milton*, ed. Nicholas McDowell and Nigel Smith (Oxford: Oxford University Press, 2012), 589–612.

his divine nature, perhaps as far from that nature as he is close to Milton's. The Son dismisses the idea of glory among men by a savage dismissal of the many:

> what the people but a herd confused,
> A miscellaneous rabble, who extol
> Things vulgar, and well weighed, scarce worth the praise?
> They praise and they admire they know not what;
> And know not whom. (3.49–53)

While he does acknowledge the few "intelligent amongst them and the wise" (3.58), this passage seems to suggest less the Son's divine love for his creation than Milton's sharp sense of superiority over the crowd, no doubt reflecting his disappointment with the English people at abandoning the Good Old Cause, as articulated so vividly at the end of *The Readie and Easie Way to Establish a Free Commonwealth* (CPW 7:462).

In excavating deep beneath the terse source passages in Matthew and Luke, Milton explores a consciousness poised between two natures—divine and human. It is significant that, at least until the end, they do not express themselves at the same time. The human consciousness slowly learns that it belongs to someone out of the ordinary run of human beings; the divine consciousness, apparently temporarily amnesiac, emerges only slowly, and completely, if it ever does, at the very end of the epic. This fits in an original and ingenious way the theology of the atonement, which held both that a man must die to expiate human sin and that only a God's death could have the miraculous potential to atone for all the sins of humankind. By the end of *Paradise Regained*, the two natures, the divine and the human, coalesce into something like the consciousness of the Theanthropos Christ.

Milton and Newton, two irascible and visionary geniuses of seventeenth-century England, play outsized roles in shaping the perspectives of succeeding generations. Milton's political writings, most notably the *Tenure of Kings and Magistrates* and *The Readie and Easie Way to Establish a Free Commonwealth*, whatever their limited effect at the time, proved inspirational to succeeding thinkers. The American Founders, among others, found inspiration in Milton's justification for the deposing of tyrants in the *Tenure* and the nascent federalism of the *Readie and Easie Way*. The *Areopagitica*, despite its apparent allowance of post-publication censorship and exception for Roman Catholics, proved foundational for defenses of freedom of speech and publication. The fact that the person on the street in the Anglophone identifies the apple as the fruit of the Tree of Knowledge owes something to the art of Lucas Cranach, Albrecht Dürer, Peter Paul Rubens, and others, but also to the ubiquity and influence of *Paradise Lost*, a work that was, along with the Bible and the *Pilgrim's Progress*, the book most likely to be found in the knapsacks of British soldiers in World War I. As for Newton, his physics still forms the commonplace conception of the interaction of matter, even after the advent of Einstein's theories of general and special relativity and the counterintuitive discoveries of quantum physics. For most applications in science and technology, Newton's laws still obtain: a body remains at rest or in motion in a straight line at constant speed unless a force acts on it; the force of an object equals its acceleration multiplied by its mass; and every action has an equal and opposite reaction. These contributions to the furniture of the modern mind, Newton's generally and Milton's for the Anglophone world, are familiar to most. In this book, I have focused instead on ways in which these two visionary figures influenced new ways of thinking about the Bible, about the freedom and responsibility of the interpreter of the Bible, and about the life of matter.

Three quarters of a century ago Geoffrey Nuttall, in *The Holy Spirit in Puritan Faith and Experience*, argued that Milton represents the full flowering and logical outcome of Protestant and, in an eminent sense, Puritan thinking about the priesthood of all believers; about the importance of being led by the Spirit; and about *sola scriptura*, the perspicuous availability in the Bible of the truths of faith necessary for salvation. Newton, as I have argued in this book, resembles Milton in all these ways, and the conclusions that he reached overlapped extensively though not entirely with Milton's. The thinking of both was influential for Unitarians, the intellectual descendants of the Socinians whose thought Milton and Newton in various ways echoed. While it would have alarmed either of them, their declarations of independence from tradition as the arbiter of theological truth were attractive to Deists, who attempted to enlist their prestige. The strong emphasis on the accessibility of the truths of faith to reason and their skepticism about mystery and sophisticated quasi-philosophical elaboration of the centrally important doctrine of the Trinity mark a stage in the unsettling of the hitherto strong European Christian consensus. This is the case despite the fervent Christianity of each, despite, that is, Milton's herculean efforts to defend the justice of the Christian God and Newton's abiding insistence on demonstrating the immediate and omnipotent hand of that God in nature and history.

Each was an unusually adept amateur in theology. In natural philosophy and mathematics, on the other hand, where Milton was an amateur, Newton was the unparalleled leader, co-inventor of the calculus, and inventor of classical physics. Despite this difference, their thinking overlapped in natural philosophy as well. Milton in *Paradise Lost* and Newton in his alchemical writing imagined a cosmos of living matter, of God's creation of a world not divided into an immaterial thinking and self-moving stuff and an essentially inactive, inert material stuff. They were harbingers of vitalist developments in the eighteenth century, and they look forward to a world in which most of us no longer think of the mind or soul as a separable, incorporeal something but as part of material nature. While Milton's poetry and Newton's physics have been recognized for their profound effects on the early modern and modern world, their lesser-known and often surprisingly related strands of thought in theology, religion, and natural philosophy foreshadowed ways in which future generations would view the world.

Milton and Newton: Matter, God, and Modernity. Stephen M. Fallon, Oxford University Press.
 DOI: 10.1093/9780198988939.003.0006

Works Cited

John Milton

Areopagitica. Edited by Ernest Sirluck. Vol. 2 of *CPW*.

The Complete Poetry and Essential Prose (*MLM*). Edited by William Kerrigan, John Rumrich, and Stephen M. Fallon. New York: Modern Library, 2007.

The Complete Prose Works of John Milton (*CPW*). Edited by Don M. Wolfe et al, 8 vols New Haven: Yale University Press, 1953–82.

The Complete Works of John Milton (*OCW*). Edited by Thomas N. Corns and Gordon Campbell. Oxford: Oxford University Press, 2009–.

De Doctrina Christiana, edited and translated by John K. Hale and J. Donald Cullington. Vol. 8 of *OCW*.

Letter to a Friend, 1633, edited by Don M. Wolfe. Vol 1 of *CPW*.

Of Education, edited by Donald C. Dorian. Vol. 2 of *CPW*.

Prolusion VII, translated by Phyllis B. Tillyard. Vol 1 of *CPW*.

The Readie and Easie Way to Establish a Free Commonwealth, 2nd ed, edited by Robert W. Ayers. Vol. 7 of *CPW*.

The Reason of Church-Government, edited by Don M. Wolfe. Vol. 1 of *CPW*.

The Shorter Poems, edited by Barbara Kiefer Lewalski and Estelle Haan. Vol. 3 of *OCW*.

A Treatise of Civil Power in Ecclesiastical Causes, edited by Robert W. Ayers. Vol. 7 of *CPW*.

Isaac Newton

Most of the manuscripts below can be found in these two databases:

The Chymistry of Isaac Newton, https://webapp1.dlib.indiana.edu/newton/.

The Newton Project, https://www.newtonproject.ox.ac.uk/.

The Chronology of Ancient Kingdoms Amended. London, 1728.

The Correspondence of Isaac Newton, edited by H. W. Turnbull. 10 vols. Cambridge: Cambridge University Press, 1959–77.

"De Gravitatione et Æquipondio Fluidorum." Cambridge University Library MS Add. 4003, Transcribed and translated in *Unpublished Scientific Papers of Isaac Newton*, edited by A. Rupert Hall and Marie Boas Hall. Cambridge, Cambridge University Press, 1962. Newton Project https://www.newtonproject.ox.ac.uk/catalogue/record/THEM00093.

"De natura acidorum." Cambridge University Library MS Add. 9597/2/18/81. https://www.newtonproject.ox.ac.uk/view/texts/diplomatic/NATP00284.

"Draft Chapters of a Treatise on the Origin of Religion and Its Corruption." Yahuda MS. 4, National Library of Israel, Jerusalem. https://www.newtonproject.ox.ac.uk/view/texts/normalized/THEM00077.

"Draft Notes on the History of the Church." Yahuda MS. 15, National Library of Israel, Jerusalem. https://www.newtonproject.ox.ac.uk/view/texts/normalized/THEM00219 and http://www.newtonproject.ox.ac.uk/view/texts/normalized/THEM00237.

"Draft Version of Queries to the *Optics*." Cambridge University Library Additional MS. 3970.3. http://www.newtonproject.ox.ac.uk/view/texts/normalized/NATP00055.

"Fitzwilliam Notebook." Fitzwilliam Museum, Cambridge. https://www.newtonproject.ox.ac.uk/view/texts/diplomatic/ALCH00069.

"Fragment on the History of Apostasy." Yahuda MS. 18, 2ᵛ, National Library of Israel, Jerusalem. https://www.newtonproject.ox.ac.uk/view/texts/normalized/THEM00061.

"An Historical Account of Two Notable Corruptions of Scripture, in a Letter to a Friend." MS. 361(4), New College Library, Oxford. https://www.newtonproject.ox.ac.uk/view/texts/normalized/THEM00264.
"Humores Minerales." MS. 1031B, Dibner Library, Smithsonian Institution, Washington D.C. https://www.newtonproject.ox.ac.uk/catalogue/record/ALCH00081.
"An Hypothesis Explaining the Properties of Light." In Vol 3 of Thomas Birch, *The History of the Royal Society*. London, 1757.
"Irenicum, or Ecclesiastical Polyty tending to Peace." Keynes MS. 3, King's College, Cambridge. https://www.newtonproject.ox.ac.uk/view/texts/normalized/THEM00003.
"Manna, Transcript of an Anonymous Alchemical Treatise, in Another Hand with Additions and Notes by Newton." Keynes MS. 33, King's College, Cambridge. https://webapp1.dlib.indiana.edu/newton/mss/norm/ALCH00022.
"Miscellaneous Notes and Extracts on the Temple, the Fathers, Prophecy, Church History, Doctrinal Issues, etc." Yahuda MS. 14, National Library of Israel, Jerusalem. http://www.newtonproject.ox.ac.uk/view/texts/normalized/THEM00057.
Newton: A Norton Critical Edition, edited by I. Bernard Cohen and Richard S. Westfall. New York: Norton, 1995.
"Notes Evidently on Newton's Own Laboratory Experiments." Cambridge University Library MS. Add. 3973. https://webapp1.dlib.indiana.edu/newton/mss/dipl/ALCH00109.
"Notes on Early Church History and the Moral Superiority of the 'barbarians' to the Romans." Yahuda MS. 39, National Library of Israel, Jerusalem. https://www.newtonproject.ox.ac.uk/view/texts/normalized/THEM00076.
Observations upon the Prophecies of Daniel, and the Apocalypse of St. John. Dublin, 1733. Eighteenth Century Collections Online, Gale doc. GALEICW0117768093.
"Of Natures Obvious Laws and Processes in Vegetation." MS. 1031B, Dibner Library, Smithsonian Institution, Washington D.C. https://webapp1.dlib.indiana.edu/newton/mss/norm/ALCH00081.
"Of the Church." Bodmer Manuscript, Fondation Martin Bodmer, Geneva, Switzerland.
Opticks: Or, A treatise of the Reflections, Refractions, Inflexions and Colours of Light. Also Two treatises of the Species and Magnitude of Curvilinear Figures. London, 1704. https://www.newtonproject.ox.ac.uk/view/texts/diplomatic/NATP00037.
"Out of Cudworth." William Andrews Clark Memorial Library, MS. shelfmark fN563Z. https://www.newtonproject.ox.ac.uk/view/texts/diplomatic/THEM00118.
"Paradoxical Questions Concerning the Morals & Actions of Athanasius & his Followers." William Andrews Clark Memorial Library, MS. shelfmark N563M3 P222. https://www.newtonproject.ox.ac.uk/view/texts/diplomatic/THEM00010.
The Principia: Mathematical Principles of Natural Philosophy. A new translation by I. Bernard Cohen and Anne Whitman, assisted by Julia Budenz. Berkeley: University of California Press, 2016.
"Prophecies Concerning Christs 2nd Coming." MS. N47 HER, James White Library, Andrews University, Berrien Springs, Michigan. https://www.newtonproject.ox.ac.uk/view/texts/normalized/THEM00088.
"Questiones quædam philosophiæ." Cambridge University Library MS. Add. 3996. https://www.newtonproject.ox.ac.uk/view/texts/normalized/THEM00092.
"A Short Schem of the True Religion." Keynes MS. 7, King's College, Cambridge. https://www.newtonproject.ox.ac.uk/view/texts/normalized/THEM00007.
Sir Isaac Newton's Theological Manuscripts, edited by Herbert McLachlan. Liverpool: At the University Press, 1950.
"*Tabula Smaragdina* [Green Tablet] of 'Hermes Trismegistus.'" Newton's translation and transcription, with notes. Keynes MS. 28, King's College, Cambridge. https://webapp1.dlib.indiana.edu/newton/mss/norm/ALCH00017.
"Theological Notebook, Part One," Keynes Ms. 2, King's College, Cambridge. https://www.newtonproject.ox.ac.uk/view/texts/normalized/THEM00180.

"Twelve Articles on Religion." Keynes MS. 8, King's College, Cambridge. https://www.newtonproject.ox.ac.uk/view/texts/normalized/THEM00008.
"Two Incomplete Treatises on Prophecy." Keynes MS. 5, King's College, Cambridge. http://www.newtonproject.ox.ac.uk/view/texts/diplomatic/THEM00005.
"Unarranged Fragments, Mostly Relating to the Dispute with Leibniz." Cambridge University Library MS. Add. 3968. https://www.newtonproject.ox.ac.uk/view/texts/diplomatic/NATP00385.
"Untitled Treatise on Revelation." Yahuda MS. 1, National Library of Israel, Jerusalem. https://www.newtonproject.ox.ac.uk/catalogue/record/THEM00044.
"Of the Woman & Two-horned Beast." Yahuda Ms. 7.2e, National Library of israel, Jerusalem. https://www.newtonproject.ox.ac.uk/view/texts/diplomatic/THEM00424.

Primary Sources

Anon. *The True Effigies of the Monster of Malmesbury, or Thomas Hobbes in his Proper Colours.* London, 1680. Early English Books Online, ProQuest doc. 2264175032.
Arius. "Thalia," translated by Aaron J. West. Fourth Century Christianity, Wisconsin Lutheran College; https://www.fourthcentury.com/arius-thalia-intro/.
Bacon, Francis. "De viis mortis." In Graham Rees, *Francis Bacon's Natural Philosophy: A New Source.* Chalfont-St. Giles: The British Society for the History of Science, 1984.
Bacon, Francis. *Novum Organum.* Vol. 4 of *The Works of Francis Bacon*, edited by James Spedding, Robert Leslie Ellis, and Douglas Denon Heath, 14 vols. London: Longman, 1859–64.
Baxter, Richard. *Of the Nature of Spirits, Especially Mans Soul.* London, 1682. Early English Books Online, ProQuest doc. 2248523326.
Best, Paul. *Mysteries Discovered, or, A Mercuriall Picture Pointing Out the Way from Babylon to the Holy City for the Good of All Such as During that Night of Generall Errour and Apostasie.* London, 1647. Early English Books Online, ProQuest doc. 2264183500.
Bible, King James Version.
Boyle, Robert. *The Excellency of Theology Compar'd with Natural Philosophy (as both are Objects of Men's Study) Discours'd of in a Letter to a Friend. To which are annex'd some Occasional Thoughts about the Excellency and Grounds of the Mechanical Hypothesis.* London, 1674. Early English Books Online, ProQuest doc. 2264186062.
Boyle, Robert. *The General History of the Air.* London, 1692. Early English Books Online, ProQuest doc. 2240859706.
Boyle, Robert. *The Origins of Forms and Qualities.* Vol. 5 in *The Works of Robert Boyle*, edited by Michael Hunter and Edward B. Davis, 14 vols. London: Pickering and Chatto, 1999.
Brewster, David. *Memoirs of the Life, Writings, and Discoveries of Sir Isaac Newton*, 2 vols. Edinburgh: T. Constable and Co., 1855.
Calvin, John. *Institutes of the Christian Religion*, edited by John T. McNeill, transated by Ford Lewis Battles. Library of Christian Classics, 2 vols. Philadelphia: Westminster Press, 1960.
Cavendish, Margaret. "The Blazing World." In *The Blazing World and Other Writings*, edited by Kate Lilley. 1992; rpt. Harmondsworth: Penguin, 2004.
Cavendish, Margaret. *Natures Picture Frawn by Fancies Pencil.* London, 1656. Early English Books Online, ProQuest doc. 2240924768.
Cavendish, Margaret. *Observations upon Experimental Philosophy*, edited by Eileen O'Neill. Cambridge Texts in the History of Philosophy. Cambridge: Cambridge University Press, 2001.
Cavendish, Margaret. *Philosophical Letters, or, Modest Reflections upon Some Opinions in Natural Philosophy Maintained by Several Famous and Learned Authors of this Age, Expressed by Way of Letters by the Thrice Noble, Illustrious, and Excellent Princess the Lady Marchioness of Newcastle.* London, 1664. Early English Books Online, ProQuest doc. 2240875599.
Charleton, Walter. *The Darknes of Atheism Dispelled by the Light of* Nature. London, 1652. Early English Books Online, ProQuest doc. 2264194095.
Charleton, Walter. *Physiologia Epicuro-Gassendo-Charletoniana, or a Fabrick of Science Natural, upon the Hypothesis of Atoms.* London, 1654. Early English Books Online, ProQuest doc. 2240941769.

Conduitt, John. "Notes on Newton's Character." Keynes Ms. 130.07, 5^r-5^v. http://www.newton-project.ox.ac.uk/view/texts/normalized/THEM00169, accessed 8/1/2019.

Conway, Anne. *The Conway Letters, The Correspondence of Anne, Viscountess Conway, Henry More and their Friends, 1642–1684*, edited by Marjorie Hope Nicolson and revised by Sarah Hutton. Oxford: Oxford University Press, 1992.

Conway, Anne. *The Principles of the Most Ancient and Modern Philosophy*, edited and translated by Allison P. Coudert and Taylore Corse. Cambridge Texts in the History of Philosophy. Cambridge: Cambridge University Press, 1996.

Conway, Anne. *The Principles of the Most Ancient and Modern Philosophy*, edited and translated by Peter Loptson. The Hague: Martinus Nijhoff, 1982.

Cowley, Abraham. "Davideis." In *Poems: Miscellanies, The Mistress, Pindarique Odes, Davideis, Verses Written on Several Occasions*, edited by A. R. Waller. Cambridge: Cambridge University Press, 1905.

Cress, Johann. *The Expiation of a Sinner*. London, 1646. Early English Books Online, ProQuest doc. 2264213504.

Cudworth, Ralph. *The True Intellectual System of the Universe*. London, 1678. Early English Books Online, ProQuest doc. 2264206863.

Dennis, John. *The Grounds of Criticism in Poetry*. London, 1704.

Descartes, René. *Discourse on the Method*. In *The Philosophical Writings of Descartes*, edited and translated by John Cottingham, Robert Stoothoff, and Dugald Murdoch, 2 vols. Cambridge: Cambridge University Press, 1984–85. I:111–51.

Descartes, René. *Le Monde*, in *The Philosophical Writings of Descartes*. I:81–98.

Descartes, René. *Principles of Philosophy*, excerpted in *The Philosophical Writings of Descartes*. I:177–291.

Donne, John. "An Anatomy of the World: The First Anniversary." In *Donne: Poetical Works*, edited by Sir Herbert Grierson. Oxford: Oxford University Press, 1933.

Donne, John. *John Donne: The Satires, Epigrams and Verse Letters*. Edited by W. Milgate. Oxford: Oxford University Press, 1967.

Edwards, John. *Some Brief Critical Remarks on Dr. Clarke's Last Papers*. London, 1714. https://newtonprojectca.files.wordpress.com/2013/06/edwards-some-remarks-on-clarkes-last-papers-postscript-1714-a4.pdf*.

Fludd, Robert. *Robert Fludd and His Philosophicall Key: Being a Transcription of the manuscript at Trinity College, Cambridge*, with an Introduction by Allen G. Debus. New York: Science History Publications, 1979.

Glisson, Francis. *Anatomia hepatis*. London, 1654. Early English Books Online, ProQuest doc. 2240883023.

Glisson, Francis. "Medical Lectures at the College of Physicians." MS Sloane 3309.

Harvey, William. *Anatomical Exercitations Concerning the Generation of Living Creatures*. London, 1653. Early English Books Online, ProQuest doc. 2240960739.

Hobbes, Thomas. *Behemoth: Or The Long Parliament*, edited from original manuscript by Ferdinand Tönnies. London: Simpkin, Marshall, and Co, 1889.

Hobbes, Thomas. *De Homine*. Vol. 4 in *The English Works of Thomas Hobbes*, edited by Sir William Molesworth, 11 vols. London: John Bohn, 1839–45.

Hobbes, Thomas. *Leviathan*, edited by C. B. Macpherson. Harmondsworth: Penguin, 1968.

Hooke, Robert. *Lampas, or, Descriptions of Some Mechanical Improvements of Lamps & Waterpoises Together With Some Other Physical and Mechanical Discoveries*. London, 1677. The Making of the Modern World, Gale doc. GALEIU0104138027.

Kant, Immanuel. *Metaphysical Foundations of Natural Science*, translated and edited by Michael Friedman. Cambridge Texts in the History of Philosophy. Cambridge: Cambridge University Press, 2004.

Kant, Immanuel. "Notes on Metaphysics." In *Notes and Fragments*, edited by Paul Guyer and translated by Curtis Bowman and Paul Guyer. *Cambridge Edition of the Works of Immanuel Kant*. Cambridge: Cambridge University Press, 2005.

Kierkegaard, Søren. *Philosophical Fragments* (1844), edited and translated by Edna H. Hong and Howard V. Hong. Princeton: Princeton University Press, 1986.

Locke, John. *An Essay Concerning Human Understanding*, edited by Pauline Phemister. Oxford: Oxford University Press, 2008.

Lucretius. *De Rerum Natura*, in *The Way Things Are*, translated by Rolfe Humphries. Bloomington, Indiana University Press, 1968.

Luther, Martin. "Preface to the Revelation of St. John (I)." Volume 6 in the *Works of Martin Luther*, edited by Andrew Spaeth et al. 6 vols. Philadelphia: Holman, 1932.

Marvell, Andrew. "On Mr. Milton's *Paradise* Lost." In *MLM*.

Melville, Herman. *Moby-Dick, or The White Whale*. 1851; rpt. Boston: St. Botolph Society, 1892.

More, Henry. *Annotations upon the Two Foregoing Treatises, Lux Orientalis, or, An Enquiry into the Opinion of the Eastern Sages Concerning the Prae-existence of Souls; and the Discourse of Truth*. London, 1682. Early English Books Online, ProQuest doc. 2264208413.

More, Henry. *The Immortality of the Soul*. Vol. 2 of *A Collection of Several Philosophical Writings*, 2 vols. London, 1662; facs. rpt. New York: Garland, 1978.

More, Henry. "The Easie, True, and Genuine Notion and Consistent Explication of the Nature of a Spirit." In *Philosophical Writings of Henry More*, edited by Florence Isabel MacKinnon. New York: Oxford University Press, 1925.

Morzkowski, Peter. *The Ecclesiastical Polity Which They Commonly Call the Agenda or The Form of the Exterior Government of the Christian Churches of Poland* (1646). In Vol. 2 of Williams, *The Polish Brethren*.

Newton, Thomas. *Paradise Lost, The Seventh Edition, with notes of Various Authors*. London, 1770.

Nolle, Heinrich. *Hermetical Physick: Or, The Right Way to Preserve, and to Restore Health*. Translated by Henry Vaughan. London: Humphrey Mosely, 1655. Early English Books Online, ProQuest doc. 2240954847.

Paracelsus (Philipp Theophrastus of Hohenheim). *Of the Nature of Things, Nine Books*, translated by J. F., M. D. London, 1650. Early English Books Online, ProQuest doc. 2264207732.

Pascal, Blaise. *Pensées*, translated by A. J. Krailshemer. 1966; rev. ed. Harmondsworth: Penguin, 1995.

Plato. *Ion* and *Phaedrus*. In *Plato: The Collected Dialogues*, edited by Edith Hamilton and Huntington Cairns. Princeton: Princeton University Press, 1961.

The Racovian Catechism. Amsterdam, 1652; Early English Books Online, ProQuest doc. 2248533629.

Raleigh, Sir Walter. *The History of the World: In Five Books*. London, 1614.

Sidney, Sir Philip, *A Defence of Poetry*. In *The Miscellaneous Prose of Sir Philip* Sidney, edited by Katherine Duncan-Jones and Jan van Dorsten. Oxford: Oxford University Press, 1973.

Spence, Joseph. *Anecdotes, Observations, and Characters, of Books and Men: Collected from the Conversation of Mr. Pope, and Other Eminent Persons of His Time*, edited by Samuel Singer. London: W. H. Carpenter, 1820.

Spinoza, Benedict de. *A Theologico-Political Treatise*, translated by R. H. M. Elwes. New York: Dover, 1951.

Stegmann, Joachim. *A Brief Inquiry Touching a Better Way than Is Commonly Made Use of to Refute Papists and Reduce Protestants to Certainty and Unity in Religion*, translated by John Biddle. London, 1653. Excerpted in Vol. 2 of Williams, *The Polish Brethren*.

Szlichtyng, Jonas. *Confession of Christian Faith Published in the name of the Polish Churches Which Profess One God and His Only-Begotten Son Jesus Christ with the Holy Spirit*. Wrocmirowa, 1642. Reprinted in Vol. 2 of Williams, *The Polish Brethren*.

Toland, John. *The Life of John Milton*. London, 1699.

Virgil. *Eclogues, Georgics, Aeneid I-VI*, with an English translation by H. Rushton Fairclough, revised by G. P. Goold. Cambridge, MA: Loeb Classical Library, 1999. https://www.loebclassics.com/view/virgil-aeneid/1916/pb_LCL063.583.xml.

Voltaire. *Letters Concerning the English Nation*. London, 1733. Internet Archive, https://archive.org/details/lettersconcernin00voltuoft.

Westminster Confession of Faith. London, 1646. https://prts.edu/wp-content/uploads/2016/12/Westminster_Confession.pdf.

Whiston, William. *Historical Memoirs of the Life and Writings of Dr. Samuel Clarke*. London, 1730.

Williams, George Huntston, ed. and trans. *The Polish Brethren: Documentation of the History & Thought of Unitarianism in the Polish-Lithuanian Commonwealth and in the Diaspora, 1601–1685*. Harvard Theological Studies XXX, 2 vols, continuously paginated. Ann Arbor: Scholars Press, 1980.

Wiszowaty, Andrew. *Rational Religion, or a Tract Concerning the Judgment of Reason to Be Used Even in Theological and Religious Controversies*. Amsterdam (?), 1685. Reprinted in Vol. 2 of Williams, *The Polish Brethren*.

Wollebius, Johannes. *Compendium Theologiae Christianae*. Geneva, 1626. In *Reformed Dogmatics*, edited and translated by John W. Beardslee III. New York: Oxford University Press, 1965.

Wordsworth, William. *The Prelude: Or Growth of a Poet's Mind*, edited by Ernest De Selincourt and Helen Darbishire. 2nd revised ed. Oxford: Oxford University Press, 1958.

Secondary Sources

Abraham, Lyndy. *Marvell and Alchemy*. Aldershot: Scolar Press, 1990.

Abraham, Lyndy. "Milton's *Paradise Lost* and 'the sounding alchymie.'" *Renaissance Studies* 12, no. 2 (1998): 261–76.

Achinstein, Sharon. "High Enterprise: Milton and the Genres of Scholarship in the Divorce Tracts." In *Scholarly Milton*, edited by Thomas Festa and Kevin J. Donovan. Liverpool: Liverpool University Press, 2019.

Aït-Touati, Frédérique. *Fictions of the Cosmos: Science and Literature in the Seventeenth Century*. Translated by Susan Emanuel. Chicago: University of Chicago Press, 2011.

Andrault, Raphaële. "What is Life? A Comparative Study of Ralph Cudworth and Nehemiah Grew." In *The Life Sciences in Early Modern Philosophy*, edited by Ohad Nachtomy and Justin E. H. Smith. Oxford: Oxford University Press, 2014.

Bauman, Michael. *Milton's Arianism*. Frankfurt: Peter Lang, 1986.

Brewster, David. *Memoirs of the Life, Writings, and Discoveries of Sir Isaac Newton*. Edinburgh: T. Constable and Co., 1855.

Budick, Sanford. *Kant and Milton*. Cambridge, MA: Harvard University Press, 2010.

Budick, Sanford. *Sublime Reciprocity in Milton, Kant and Wordsworth*. Edinburgh: Edinburgh University Press, 2025.

Burgess, Thomas. *Milton Not the Author of the Lately-discovered Arian Work De Doctrina*. London: Thomas Brettell, 1829.

Campbell, Gordon, and Thomas N. Corns. *John Milton: Life, Work, and Thought*. Oxford: Oxford University Press, 2010.

Campbell, Gordon, Thomas N. Corns, John K. Hale, and Fiona J. Tweedie. *Milton and the Manuscript of "De Doctrina Christiana."* Oxford: Oxford University Press, 2010.

Chang, Ku-ming (Kevin). "Alchemy as Studies of Life and Matter: Reconsidering the Place of Vitalism is Early Modern Chymistry." *Isis* 102, no. 2 (2011): 322–29.

Chapman, Alison A. "'To All Baptized': The Watery Dissent of *Paradise Regained*." *Milton Studies* 65 (2023): 276–99.

Cohen, I. Bernard. "A Guide to Newton's *Principia*." In Isaac Newton, *The Principia: Mathematical Principles of Natural Philosophy*, edited by I. Bernard Cohen and Anne Whitman. Berkeley: University of California Press, 2016.

Clawson, James M., and Hugh F. Wilson. "*De Doctrina Christiana* and Milton's Canonical Works: Revisiting the Authorship Question." *Renaissance and Reformation* 44, no. 3 (2021): 151–98.

Danielson, Dennis. "Astronomy." In *Milton in Context*, edited by Stephen B. Dobranski. Cambridge: Cambridge University Press, 2010.

Debus, Allen. "Chemistry and the Quest for a Material Spirit of Life in the Seventeenth Century." In *Spiritus: IVth Colloquio Internazionale*, edited by M. Fattori and M. Bianchi. Rome: Edizioni dell'Ateneo, 1984.

Debus, Allen. "The Paracelsian Aerial Niter." *Isis* 55, no. 1 (1964): 43–61.

Dobbs, Betty Jo Teeter. *The Foundations of Newton's Alchemy, or, The Hunting of the Greene Lyon*. Cambridge: Cambridge University Press, 1975.

Dobbs, Betty Jo Teeter. *The Janus Faces of Genius: The Role of Alchemy in Newton's Thought*. Cambridge: Cambridge University Press, 1991.

Dobranski, Stephen B. "Licensing Milton's Heresy." In *Milton and Heresy*, edited by Stephen B. Dobranski and John Rumrich. Cambridge: Cambridge University Press, 1998.

Donnelly, Phillip J. "'Matter' versus Body: The Character of Milton's Monism." *Milton Quarterly* 33, no. 3 (1999): 79–85.

Dzelzainis, Martin. "Milton and Antitrinitarianism." In *Milton & Toleration*, edited by Sharon Achinstein and Elizabeth Sauer. Oxford: Oxford University Press, 2007.

Fallon, Samuel. "Milton's Strange God: Theology and Narrative Form in *Paradise Lost*." *ELH: English Literary History* 79, no. 1 (2012): 33–57.

Fallon, Stephen M. "Imagining Animist Materialism in *Paradise Lost* and *the Blazing World*." In *Margaret Cavendish and John* Milton, edited by Lara A. Dodds, Lisa Walters, and Ann Baynes Coiro. Oxford: Oxford University Press, forthcoming.

Fallon, Stephen M. *Milton among the Philosophers: Poetry and Materialism in Seventeenth-Century England*. Ithaca: Cornell University Press, 1991.

Fallon, Stephen M. "Milton's Arminianism and the Authorship of *De Doctrina Christiana*." *Texas Studies in Literature and Language* 41, no. 2 (1999): 103–27.

Fallon, Stephen M. *Milton's Peculiar Grace: Self-representation and Authority*. Ithaca: Cornell University Press, 2007.

Fallon, Stephen M. "Narrative and Theodicy in *Paradise Lost*." *Milton Studies* 61 (2019): 40–64.

Fallon, Stephen M. "'To Act or Not': Milton's Conception of Divine Freedom." *Journal of the History of Ideas* 49, no. 3 (1988): 425–49.

Feingold, Mordechai. "Isaac Newton, Heretic? Some Eighteenth-Century Perception." In *Reading Newton in Early Modern Europe*, edited by Elizabethanne Boran and Mordechai Feingold. Leiden: Brill, 2017.

Force, James E. "The God of Abraham and Isaac (Newton)." In *The Books of Nature and Scripture*, edited by J. E. Force and R. H. Popkin. Dordrecht: Kluwer Academic Publishing, 1994.

Force, James E. "The Newtonians and Deism." In *Essays on the Context, Nature, and Influence of Isaac Newton's Theology*, edited by James E. Force and Richard H. Popkin. Dordrecht: Kluwer Academic Publishing, 1990.

Force, James E. "Newton's God of Dominion: The Unity of Newton's Theological, Scientific, and Political Thought." In *Essays on the Context, Nature, and Influence of Isaac Newton's Theology*, edited by James E. Force and Richard H. Popkin. Dordrecht: Kluwer, 1990.

Force, James E., and Sarah Hutton, eds. *Newton and Newtonianism: New Studies*. Dordrecht: Kluwer, 2004.

Force, James E., and Richard H. Popkin, eds. *The Books of Nature and Scripture: Recent Essays on Natural Philosophy, Theology, and Biblical Criticism in the Netherlands of Spinoza's Time and the British Isles of Newton's Time*. Dordrecht: Kluwer, 1994.

Force, James E., and Richard H. Popkin. *Essays on the Context, Nature, and Influence of Isaac Newton's Theology*. Dordrecht: Kluwer, 1990.

Gabbey, Alan. "Philosophia Cartesiana Triumphata: Henry More (1646-1671)." In *Problems of Cartesianism*, edited byThomas M. Lennon, John M. Nicholas, and John W. Davis. Montreal: McGill-Queens University Press, 1982.

Giglione, Guido. "The Genesis of Francis Glisson's Philosophy of Life." Ph.D. dissertation, Johns Hopkins University, 2002.

Goldish, Matt. "Newton's 'Of the Church: Its Contents and Implications." In *Newton and Religion: Context, Nature, and Influence*, edited by James E. Force and Richard H. Popkin. Dordrecht: Kluwer Academic Publishers, 1999.

Hanson, R. P. C. *The Search for the Christian Doctrine of God: The Arian Controversy 318–381.* London: T & T Clark, 2005.
Harrison, John. *The Library of Isaac Newton.* Cambridge: Cambridge University Press, 1978.
Harrison, Peter. "'I Believe Because it is Absurd': The Enlightenment Invention of Tertullian's Credo." *Church History* 86, no. 2 (2017): 339–364.
Harrison, Timothy M. *Coming To: Consciousness and Natality in Early Modern England.* Chicago: University of Chicago Press, 2020.
Harrison, Timothy M. "Every Living Thing": Life and Endeavor in Milton's *Paradise Lost.*" *Representations* 167, no. 1 (2024): 33–63.
Haskin, Dayton. *Milton and the Burden of Interpretation.* Philadelphia, University of Pennsylvania Press, 1994.
Henry, John. "The Matter of Souls: Medical Theory and Theology in Seventeenth-Century England." In *The Medical Revolution of the Seventeenth Century*, edited by R. K. French and Andrew Wear. Cambridge: Cambridge University Press, 1989.
Henry, John. "Newton, the Sensorium of God, and the Cause of Gravity." *Science in Context* 33, no. 3 (2020): 329–51; https://doi.org/10.1017/S0269889721000077.
Henry, John. "Occult Qualities and the Experimental Philosophy: Active Principles in Pre-Newtonian Matter Theory." *History of Science* 24, no. 4 (1986): 335–81.
Hequembourg, Stephen. "The Poetics of Materialism in Cavendish and Milton." *SEL: Studies in English Literature 1500–1900* 54, no. 1 (2014): 173–92.
Hill, Christopher. *Milton and the English Revolution.* New York: Viking, 1977.
Hoxby, Blair, and Ann Baynes Coiro, eds. *Milton in the Long Restoration.* Oxford: Oxford University Press, 2016.
Hume, Patrick. *Annotations on Milton's "Paradise Lost."* London, 1695. Early English Books Online, ProQuest doc. 2240982831.
Hunter, William B., J. H. Adamson, and C. A. Patrides. *Bright Essence: Studies in Milton's Theology.* Salt Lake City: University of Utah Press, 1971.
Iliffe, Rob. "Abstract Considerations: Disciplines and the Incoherence of Newton's Natural Philosophy." *Studies in the History and Philosophy of Science* 35, no. 3 (2004): 427–54.
Iliffe, Rob. *Priest of Nature: The Religious Worlds of Isaac Newton.* Oxford: Oxford University Press, 2017.
Jonas, Hans. *The Phenomenon of Life: Toward a Philosophical Biology.* Chicago: University of Chicago Press, 1966.
Keleher, Sarah. "'So all was cleared': Corrupted Fancy and Purgative Tears in *Paradise Lost.*" In *Milton in Strasbourg*, edited by Christophe Tournu, John Hale, and Neil Forsyth. Brussels: Peter Lang, 2022.
Kerr, Jason. "Felbinger Not the Author of *De Doctrina Christiana*, Currently Attributed to Milton." *Notes & Queries* 72, no. 1 (2025): 65–68.
Kerr, Jason. *Milton's Theological Process: Reading "De Doctrina Christiana"and "Paradise Lost."* Oxford: Oxford University Press, 2024.
Kerrigan, William. *The Prophetic Milton.* Charlottesville, University Press of Virginia, 1974.
Koyré, Alexander. *From the Closed World to the Infinite Universe.* Baltimore: Johns Hopkins University Press, 1968.
Leonard, John. *Faithful Labourers: A Reception History of Paradise Lost, 1667–1970*, 2 vols. Oxford: Oxford University Press, 2013.
Leslie, Charles. "Preface to *The History of Sin and Heresy.*" In John T. Shawcross, *Milton: The Critical Heritage.* London: Routledge & Kegan Paul, 1970.
Levitin, Dmitri. *Ancient Wisdom in the Age of the New Science: Histories of Philosophy in England, c. 1640–1700.* Cambridge: Cambridge University Press, 2015.
Levitin, Dmitri. *The Kingdom of Darkness: Bayle, Newton, and the Emancipation of the European Mind from Philosophy.* Cambridge: Cambridge University Press, 2022.
Lewalski, Barbara. *The Life of John Milton: A Critical Biography.* Oxford: Blackwell, 2000.
Lewis, C. S. *A Preface to "Paradise Lost."* New York: Oxford University Press, 1942.
Lieb, Michael. "Milton and 'Arianism.'" *Religion & Literature* 32, no. 2 (2000): 197–220.

Lieb, Michael. "Milton and the Socinian Heresy." In *Milton and the Grounds of Contention*, edited by Mark R. Kelly, Michael Lieb, and John T. Shawcross. Pittsburgh: Duquesne University Press, 2003.

Lieb, Michael. *Theological Milton: Deity, Discourse and Heresy in the Miltonic Canon*. Pittsburgh: Duquesne University Press, 2006.

Lim, Paul C. H. *Mystery Unveiled: The Crisis of the Trinity in Early Modern England*. Oxford: Oxford University Press, 2012.

Loewenstein, David. *Treacherous Faith: The Specter of Heresy in Early Modern Literature and Culture*. Oxford: Oxford University Press, 2013.

MacCallum, Hugh. *Milton and the Sons of God: The Divine Image in Milton's Epic Poetry*. Toronto: University of Toronto Press, 1986.

Marion, Jean Luc. *On Descartes' Passive Thought: The Myth of Cartesian Dualism*. Chicago: University of Chicago Press, 2018.

Mandelbrote, Scott. "'A Duty of the Greatest Moment': Isaac Newton and the Writing of Biblical Criticism." *The British Journal for the History of Science* 26, no. 3 (1993): 281–302.

Manuel, Frank. *The Religion of Isaac* Newton. Oxford: Clarendon Press, 1974.

Marjara, Harinder Singh. *Contemplation of Created Things: Science in "Paradise Lost."* Toronto: University of Toronto Press, 1992.

Martin, Catherine Gimelli. "'What If the Sun Be Center to the World': Milton's Epistemology, Cosmology, and Paradise of Fools Reconsidered." *Modern Philology* 99, no. 2 (2001): 231–65.

Masson, David. *The Life of John Milton: Narrated in Connexion with the Political, Ecclesiastical and Literary History of His Time*, 7 vols. London, 1881–94; rpt. Gloucester, MA: Peter Smith, 1965.

McDowell, Nicholas. *Poet of Revolution: The Making of John Milton*. Princeton: Princeton University Press, 2020.

McGuire, J. E., and P. M. Rattansi. "Newton and the Pipes of Pan." *Notes and Records of the Royal Society of London* 21, no. 2 (1966): 108–43.

McMullin, Ernan. *Newton on Matter and Activity*. Notre Dame, IN: University of Notre Dame Press, 1978.

Mintz, Samuel. *The Hunting of Leviathan: Seventeenth-Century Reactions to the Materialism and Moral Philosophy of Thomas Hobbes*. Cambridge: Cambridge University Press, 1962.

Mortimer, Sarah. *Reason and Religion in the English Revolution: The Challenge of Socinianism*. Cambridge: Cambridge University Press, 2010.

Mueller, Janel. "Milton on Heresy." In *Milton and Heresy*, edited by Stephen B. Dobranski and John Rumrich. Cambridge: Cambridge University Press, 1998.

Newman, William R. *Newton the Alchemist: Science, Enigma, and the Quest for Nature's "Secret Fire."* Princeton: Princeton University Press, 2019.

Newman, William R. "Newton's Reputation as an Alchemist and the Tradition of Chymiatria." In *Reading Newton in Early Modern Europe*, edited by Elizabethanne Boran and Mordechai Feingold. Leiden: Brill, 2017.

Nicholls, Charlotte. "Body Out of Spirit: Medical Science and the Creation of Living Soul in *Paradise Lost*." *Milton Studies* 57 (2016): 119–53.

Nunez, Tyke. "Kant on Plants: Self-Activity, Representations, and the Analogy with Life." Ann Arbor, MI: Michigan Publishing, University of Michigan Library, *Philosopher's Imprint* 21, no. 11, (2021). https://quod.lib.umich.edu/p/phimp/3521354.0021.011/1.

Oxford Dictionary of National Biography

Parker, Willliam Riley. *Milton: A Biography*, 2 vols, continuously paginated. Oxford: Oxford University Press, 1968.

Pasnau, Robert. *Metaphysical Themes: 1274–1671*. Oxford: Oxford University Press, 2011.

Popkin, Richard. "Newton's Biblical Theology and His Theological Physics." In *Newton's Scientific and Philosophical Legacy*, edited by P. B. Scheuer and G. Debrock. Dordrecht: Kluwer, 1988.

Poole, William. *The World Makers: Scientists of the Restoration and the Search for the Origins of the Earth*. Oxford: Peter Lang, 2010.

Principe, Lawrence. *The Secrets of Alchemy*. Chicago: University of Chicago Press, 2013.
Richardson, Jonathan, Father and Son. *Explanatory Notes and Remarks on Milton's Paradise Lost*. Excerpted in John T. Shawcross, *Milton: The Critical Heritage, Volume II, 1732–1801*. 1972; rpt. London: Routledge, 1995.
Rogers, John. *The Matter of Revolution: Science, Poetry, and Politics in the Age of Milton*. Ithaca: Cornell University Press, 1996.
Rogers, John. "Milton and the Heretical Priesthood of Christ." In *Heresy, Literature and Politics in Early Modern England*, edited by David Loewenstein and John Marshall. Cambridge: Cambridge University Press, 2006.
Rogers, John. "Newton's Arian Epistemology and the Cosmogony of *Paradise Lost*." *ELH: English Literary History* 86, no. 1 (2019): 77–106.
Rogers, John. "*Paradise Regained* and the Memory of *Paradise Lost*." In *The Oxford Handbook of Milton*, edited by Nicholas McDowell and Nigel Smith. Oxford: Oxford University Press, 2012.
Rumrich, John. "Milton's Arianism: Why It Matters." In *Milton and Heresy*, edited by Stephen B. Dobranski and John Rumrich. Cambridge: Cambridge University Press, 1998.
Rumrich, John. "Milton's God and the Matter of Chaos." *Publications of the Modern Language Association* 110, no. 5 (1995): 1035–46.
Rumrich, John. "Milton's *Theanthropos*: The Body of Christ in *Paradise Regained*." *Milton Studies* 42 (2003): 50–67.
Rumrich, John. *Milton Unbound: Controversy and Interpretation*. Cambridge: Cambridge University Press, 1996.
Sarasohn, Lisa T. *The Natural Philosophy of Margaret Cavendish: Reason and Fancy during the Scientific Revolution*. Baltimore: Johns Hopkins University Press, 2010.
Saurat, Denis. *Milton, Man and Thinker*. New York: Dial Press, 1925.
Schliesser, Eric. "Newton and Spinoza: On Motion and Matter (and God, of Course)." *The Southern Journal of Philosophy* 50, no. 3 (2012): 436–58.
Smith, Nigel. "'And if God was one of us': Paul Best, John Biddle, and Anti-Trinitarian Heresy in Seventeenth-century England." In *Heresy, Literature and Politics in Early Modern English Culture*, edited by David Loewenstein and John Marshall. Cambridge: Cambridge University Press, 2006.
Snobelen, Stephen D. "Isaac Newton, Heretic: The Strategies of a Nicodemite." *British Journal for the History of Science* 32, no. 4 (1999): 381–419.
Snobelen, Stephen D. "Isaac Newton, Socinianism, and 'the One Supreme God.'" In *Socinianism and Arminianism: Antitrinitarians, Calvinists, and Cultural Exchange in Seventeenth-Century Europe*, edited by Martin Mulsow and John Rohls. Leiden: Brill, 2005.
Snobelen, Stephen D. "Lust, Pride, and Ambition: Isaac Newton and the Devil." In *Newton and Newtonianism: New Studies*, edited by James E. Force and Sarah Hutton. Dordrecht: Kluwer, 2004.
Snobelen, Stephen D. "On Reading Isaac Newton's *Principia* in the Eighteenth Century." *Endeavour* 22, no. 4 (1998): 159–63.
Snobelen, Stephen D. "To Discourse of God: Isaac Newton's Heterodox Theology and His Natural Philosophy." In *Science and Dissent in England, 1688–1945*, edited by Paul Wood. London: Routledge, 2004.
Snobelen, Stephen D. "'A Time and Times and the Dividing of Time': Isaac Newton, the Apocalypse, and 2060 A. D." *Canadian Journal of History* 38, no. 3 (2003): 537–51.
Spence, Joseph. *Anecdotes, Observations, and Characters, of Books and Men: Collected from the Conversation of Mr. Pope, and Other Eminent Persons of His Time*, edited by Samuel Singer. London: W. H. Carpenter, 1820.
Stavely, Keith. *Puritan Legacies: Paradise Lost and the New England Tradition, 1630–1890*. Ithaca: Cornell University Press, 1987.
Stewart, Dustin. "Paradise Regained on Socinian Time." *Religion and Literature* 45, no. 1 (2013): 180–192.
Sugimura, Noel. *Matter of Glorious Trial: Spiritual and Material Substance in "Paradise Lost."* New Haven, CT: Yale University Press, 2009.

Sugimura, Noel. "Milton and Matter." Oxford Handbook Topics in Literature (online edn, Oxford Academic, 16 December 2013); https://doi.org/10.1093/oxfordhb/9780199935338.013.11.

Svendsen, Kester. "Milton and Medical Lore." *Bulletin of the History of Medicine* 13, no. 2 (1943): 158–84.

Theophilus (Anon.). *Gentlemen's Magazine*, March 1738, excerpted in John T. Shawcross, *Milton: The Critical Heritage, Volume II, 1732–1801*. 1972. Reprint, London: Routledge, 1995.

Thomson, Ann. *Bodies of Thought: Science, Religion, and the Soul in the Early Enlightenment*. Oxford: Oxford University Press, 2008.

Trubowitz, Rachel. "Reading Milton and Newton in the Radical Reformation: Poetry, Mathematics, and Religion." *ELH: English Literary History* 84, no. 1 (2017): 33–62.

Urban, David V. "The Increasing Distance between *De Doctrina Christiana* and Milton's Poetry: An Answer to John K. Hale." *Connotations* 32, no. 1 (2023): 1–10.

Webster, Charles. *From Paracelsus to Newton: Magic and the Making of Modern Science*. Cambridge: Cambridge University Press, 1982.

Webster, Erin. "Milton's Pandæmonium and the Infinitesimal Calculus." *English Literary Renaissance* 45, no. 3 (2015): 425–58.

Webster, Erin. *The Curious Eye: Optics and Imaginative Literature in Seventeenth-Century England*. Oxford: Oxford University Press, 2020.

Westfall, Richard. "Isaac Newton's *Theologiae Gentilis Origines Philosophicae*." In *The Secular Mind: Transformation of Faith in Modern Europe*, edited by W. Warren Wagar. New York: Holmes & Meier, 1982.

Westfall, Richard. *Never at Rest: A Biography of Isaac Newton*. Cambridge: Cambridge University Press, 1980.

Westfall, Richard. "Newton's Scientific Personality." *Journal of the History of Ideas* 48, no. 4 (1987): 551–70.

Wiles, Maurice. *The Archetypal Heresy: Arianism through the Centuries*. Oxford: Oxford University Press, 2001.

Wilson, Catherine. *Epicureanism at the Origins of Modernity*. Oxford: Oxford University Press, 2008.

Windsor, Timothy. "The War in Heaven: Milton's Apotheosis of Epic Catabasis." *The Journal of Religious History* 3, no. 1 (2017): 100–21.

Yolton, John. *Thinking Matter: Materialism in Eighteenth-Century Britain*. Minneapolis: University of Minnesota Press, 1983.

Index

For the benefit of digital users, indexed terms that span two pages (e.g., 52–53) may, on occasion, appear on only one of those pages.